DEVELOPING MULTI-TENANT APPLICATIONS FOR THE CLOUD ON WINDOWS AZURE™

3RD EDITION

Developing Multi-tenant Applications for the Cloud on Windows Azure™

3rd Edition

Dominic Betts
Alex Homer
Alejandro Jezierski
Masashi Narumoto
Hanz Zhang

978-1-62114-022-1

Contents

Foreword: Bill Hilf   xi

Preface   **xiii**
    Who This Guide Is For   xiii
    Why This Guide Is Pertinent Now   xiv
    How This Guide Is Structured   xiv
    What You Need to Use the Code   xv
    Where to Go for More Information   xvi
    Who's Who   xvi

Acknowledgments   **xix**
    Acknowledgements of Contributors to the Third Edition   xxi

The Tailspin Scenario   1
    The Tailspin Company   1
        Tailspin's Strategy   1
        The Surveys Application   2
        Tailspin's Goals and Concerns   3
    The Surveys Application Architecture   5
    More Information   7

Hosting a Multi-Tenant Application on Windows Azure   9
    Goals and Requirements   9
        The Tenant's Perspective   9
        The Provider's Perspective   10
    Single Tenant vs. Multiple Tenant   11
    Multi-Tenancy Architecture in Windows Azure   13
    Selecting a Single-Tenant or Multi-Tenant Architecture   14
        Architectural Considerations   14
            Application Stability   14
            Making the Application Scalable   15
            Resource Limitations and Throttling   18
            Geo-location   19

Service Level Agreements   19
The Legal and Regulatory Environment   19
Handling Authentication and Authorization   19
The Command Query Responsibility Segregation (CQRS)
Pattern   20
Application Life Cycle Management Considerations   20
Maintaining the Code Base   20
Handling Application Updates   21
Monitoring the Application   21
Using Third-Party Components   21
Provisioning for Trials and New Subscribers   22
Customizing the Application   22
Customizing the Application by Tenant   22
URLs to Access the Application   23
Financial Considerations   24
Billing Subscribers   24
Managing Application Costs   26
Engineering Costs   26
More Information   27

Choosing a Multi-Tenant Data Architecture   29
Storing Data in Windows Azure Applications   29
Windows Azure Table Storage   29
Windows Azure Blob Storage   30
Windows Azure SQL Database   30
Other Storage Options   31
Storage Availability   31
Multi-Tenant Data Architectures   32
Partitioning to Isolate Tenant Data   32
Shared Access Signatures   35
Data Architecture Extensibility   36
Data Architecture Scalability   38
An Example   39
Option 1 — Using a Single Table   40
Option 2 — Table per Tenant   40
Option 3 — Table per Base Entity Type   40
Option 4 — Table per Entity Type   41
Option 5 — Table per Entity Type per Tenant   41
Comparing the Options   42
Goals and Requirements   42
Isolation of Tenants' Data   42
Application Scalability   43
Extensibility   43
Paging through Survey Results   43
Exporting Survey Data to SQL Database for Analysis   43

Overview of the Solution   44
    Storage Accounts   44
    The Surveys Data Model   44
        Storing Survey Definitions   45
        Storing Tenant Data   49
        Storing Survey Answers   50
        Storing Survey Answer Summaries   51
    Comparing Paging Solutions   52
        Paging with Table Storage   52
        Paging with Blob Storage   53
        Comparing the Solutions   53
    The SQL Database Design   53
Inside the Implementation   55
    The Data Store Classes   55
        SurveyStore Class   55
        SurveyAnswerStore Class   55
        SurveyAnswersSummaryStore Class   55
        SurveySqlStore Class   55
        SurveyTransferStore Class   55
        TenantStore Class   56
    Accessing Custom Data Associated with a Survey   56
        Defining a Tenant's Custom Fields   56
        Writing Custom Fields to the Surveys Table   57
        Reading Custom Fields from the Surveys Table   61
    Implementing Paging   62
    Implementing the Data Export   64
    Displaying Questions   66
    Displaying the Summary Statistics   68
More Information   69

Partitioning Multi-Tenant Applications   71
Partitioning a Windows Azure Application   71
    Partitioning Web and Worker Roles   73
        Identifying the Tenant in a Web Role   74
        Identifying the Tenant in a Worker Role   77
    Partitioning Queues   78
    Partitioning Caches   80
Goals and Requirements   81
    Isolation   81
    Scalability   81
    Accessing the Surveys Application   82
    Premium Subscriptions   82
    Designing Surveys   83

Overview of the Solution   84
    Partitioning Queues and Worker Roles   84
    Tenant Isolation in Web Roles   84
    DNS Names, Certificates, and SSL in the Surveys Application   85
        https://tailspin.cloudapp.net   86
        http://tailspin.cloudapp.net   87
        Accessing Tailspin Surveys in Different Geographic
        Regions   87
    Maintaining Session State   87
    Isolating Cached Tenant Data   89
Inside the Implementation   90
    Prioritizing Work in a Worker Role   90
        The BatchMultipleQueueHandler and the Related Classes   92
    Using MVC Routing Tables   97
    Web Roles in Tailspin Surveys   100
    Implementing Session Management   102
        Configuring a Cache in Windows Azure Caching   106
        Configuring the Session State Provider in the TailSpin.Web
        Application   107
    Caching Frequently Used Data   108
More Information   111

Maximizing Availability, Scalability, and Elasticity   113
Maximizing Availability in Multi-Tenant Applications   113
Maximizing Scalability in Multi-Tenant Applications   114
    Caching   115
    SQL Database Federation   115
    Shared Access Signatures   116
    Content Delivery Network   116
Implementing Elasticity in Multi-Tenant Applications   116
Scaling Windows Azure Applications with Worker Roles   117
    Example Scenarios for Worker Roles   118
        Triggers for Background Tasks   119
        Execution Model   120
        The MapReduce Algorithm   123
Goals and Requirements   123
    Performance and Scalability when Saving Survey
    Response Data   123
    Summary Statistics   124
    Geo-location in the Surveys Application   125
    Making the Surveys Application Elastic   126
    Scalability   126

Overview of the Solution   127
  Options for Saving Survey Responses   127
    Writing Directly to Storage   127
    Using the Delayed Write Pattern   128
    Comparing the Options   132
  Options for Generating Summary Statistics   137
    Scaling out the Generate Summary Statistics Task   139
  Using Windows Azure Caching   139
  Using the Content Delivery Network   140
    Setting the Access Control for the BLOB Containers   141
    Configuring the CDN and Storing the Content   141
    Configuring URLs to Access the Content   142
    Setting the Caching Policy   143
  Hosting Tailspin Surveys in Multiple Locations   144
    Synchronizing Survey Statistics   145
  Autoscaling and Tailspin Surveys   147
Inside the Implementation   147
  Saving the Survey Response Data Asynchronously   148
  Calculating the Summary Statistics   150
  Pessimistic and Optimistic Concurrency Handling   154
More Information   156

Securing Multi-Tenant Applications   157
Protecting Users' Data in Multi-Tenant Applications   157
  Authentication   157
  Authorization   158
  Protecting Sensitive Data   158
    Splitting Sensitive Data across Multiple Subscriptions   160
  Using Shared Access Signatures   161
Goals and Requirements   163
  Authentication and Authorization   163
  Privacy   163
Overview of the Solution   164
  Identity Scenarios in the Surveys Application   164
    Integrating a Subscribers Own Identity Mechanism   164
    Providing an Identity Mechanism for Small Organizations   165
    Integrating with Social Identity Providers   166
    Windows Azure Access Control Service and Windows
    Azure Active Directory   167
  Configuring Identity Federation for Tenants   168
  Encrypting Session Tokens in a Windows Azure Application   169
Inside the Implementation   169
  Using Windows Identity Foundation   170
  Protecting Session Tokens in Windows Azure   174
More Information   175

Managing and Monitoring Multi-Tenant Applications 177
  ALM Considerations for Multi-Tenant Applications 177
    Goals and Requirements 177
    Overview of the Solution 179
      Testing Strategies 179
      Stress Testing and Performance Tuning 181
      Application Deployment and Update Strategies 182
      Application Management Strategies 182
      Application Monitoring Strategies 185
    Inside the Implementation 186
      Unit Testing 186
      Testing Worker Roles 191
      Testing Multi-Tenant Features and Tenant Isolation 193
      Performance and Stress Testing 194
      Managing the Surveys Application 197
      Monitoring the Surveys Application 198
  ISV Considerations for Multi-Tenant Applications 199
    Goals and Requirements 199
    Overview of the Solution 200
      Onboarding for Trials and New Subscribers 200
      Configuring Subscribers 201
      Supporting Per Tenant Customization 201
      Financial Goals and Billing Subscribers 202
    Inside the Implementation 204
      Onboarding for Trials and New Subscribers 204
      Customizing the Surveys Application for Each Subscriber 209
      Billing Subscribers in the Surveys Application 212
  More Information 213

Glossary 215

Index 219

Foreword: Bill Hilf

Whether you regard it as evolution or revolution, there's no doubt that the cloud is changing the way our industry works. It presents us with exciting new opportunities for implementing modern applications. It's also changing the way we view operating systems, data storage, development languages, operations and IT infrastructure. I'm proud, in my own career, to have had the opportunity to play a part in the evolution of Microsoft's cloud platform, Windows Azure.

In addition to rich platform services for building new applications, Windows Azure provides Infrastructure as a Service (IaaS) support for both Windows Server and Linux operating systems, and simple automated integration with a wide range of open source software such as databases, blogs, forums, and more; which reinforces just how flexible and powerful Windows Azure really is. The package of highly integrated services, features, options, and manageability that it offers allows you to create almost any kind of application in the cloud; and get great performance and reliability built right in. No matter whether it's .NET, node.js, PHP, Python, or Java—you bring your designs and your applications and we provide the environment, allowing you to focus on your apps and not the infrastructure.

One of the areas where Windows Azure really scores is performance and reliability. Learning from our many years of building mission critical enterprise software and also running huge public online services, we've built an enterprise-ready infrastructure with datacenters across the globe so that you can deploy what you need, where you need it, and give your customers the best possible experience.

Your customers' concerns include a whole range of additional factors such as security, privacy, corporate presence, and regulatory requirements. This guide, from the patterns & practices team here at Microsoft, will help you to think about how you address these concerns, how Windows Azure can help you to meet your requirements, and how you can get the most benefit from our cloud platform and services. Based on a fictitious company that needs to build a real-world, multi-tenant application, the guide walks through the decision making, planning, design, and implementation of Tailspin's Surveys application. It also discusses how Tailspin tests and deploys the application, and manages and monitors it as it runs.

The team that created this guide worked closely with the Windows Azure development team to ensure that their guidance is accurate, useful, and up to date. Yes, they discuss many different options so that you get to see the range and flexibility of Windows Azure, but they also help you to choose what will best suit the specific needs of your own applications. You want solid guidance, good practice advice, working examples, hands-on labs, and plenty links to help you find out more? If so, you are already reading the right book! I hope you enjoy it.

Bill Hilf
General Manager
Windows Azure Product Marketing
Microsoft Corporation

Preface

How can a company create an application that has truly global reach and that can scale rapidly to meet sudden, massive spikes in demand? Historically, companies had to invest in building an infrastructure capable of supporting such an application themselves and, typically, only large companies would have the resources available to risk such an enterprise. Building and managing this kind of infrastructure is not cheap, especially because you have to plan for peak demand, which often means that much of the capacity sits idle for much of the time. The cloud has changed the rules of the game. By making the infrastructure available on a "pay as you go" basis, creating a massively scalable, global application is within the reach of both large and small companies.

The cloud platform provides you with access to capacity on demand, fault tolerance, distributed computing, data centers located around the globe, and the capability to integrate with other platforms. Someone else is responsible for managing and maintaining the entire infrastructure, and you only pay for the resources that you use in each billing period. You can focus on using your core domain expertise to build and then deploy your application to the data center or data centers closest to the people who use it. You can then monitor your applications, and scale up or scale back as and when the capacity is required.

Yes, by locating your applications in the cloud you're giving up some control and autonomy, but you're also going to benefit from reduced costs, increased flexibility, and scalable computation and storage. This guide shows you how to do this.

Who This Guide Is For

This guide is the second volume in a series about Windows Azure. Volume 1, *Moving Applications to the Cloud*, discusses the hosting options, cost model, and application life cycle management for cloud-based applications; and describes several scenarios for migrating an existing ASP.NET application to the cloud. This guide demonstrates how you can create from scratch a multi-tenant, Software as a Service (SaaS) application to run in the cloud by using the latest versions of the Windows Azure tools and the latest features of Windows Azure.

The guide is intended for any architect, developer, or information technology (IT) professional who designs, builds, or operates applications and services that run on or interact with the cloud. Although applications do not need to be based on the Microsoft Windows operating system to work in Windows Azure, or be written using a .NET language, this guide is written for people who work with Windows based systems. You should be familiar with the Microsoft .NET Framework, Microsoft Visual Studio development system, ASP.NET MVC, and Microsoft Visual C#.

WHY THIS GUIDE IS PERTINENT NOW

In general, the cloud has become a viable option for making your applications accessible to a broad set of customers. In particular, Windows Azure now has in place a complete set of tools for developers and IT professionals. Developers can use the tools they already know, such as Visual Studio, to write their applications for the cloud. In addition, Windows Azure SDK includes a storage emulator and a compute emulator that developers can use to locally write, test, and debug their applications before they deploy them to the cloud. There are also tools and an API to manage your Windows Azure accounts. This guide shows you how to use all these tools in the context of a common scenario—how to develop a brand new, multi-tenant, SaaS application for Windows Azure.

HOW THIS GUIDE IS STRUCTURED

Here is the tube map for the guide:How This Book Is Structured

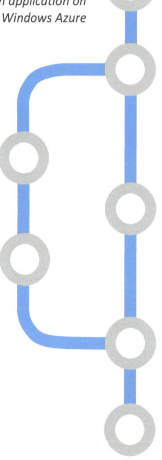

The Tailspin Scenario
Motivations, constraints, and goals of a SaaS ISV building an application on Windows Azure

Hosting a Multi-tenant Application on Windows Azure
Selecting a single or a multi-tenant architecture, stability, scalability, SLAs, authentication, ALM, monitoring, customization

Partitioning Multi-tenant Applications
Partitioning for tenants, session state management, caching, using MVC

Maximizing Availability, Scalability, and Elasticity
Geo-location, CDN, asynchronous execution, autoscaling roles

Choosing a Multi-tenant Data Architecture
Data models, partitioning, extensibility and scalability. Using Windows Azure SQL Database, Windows Azure blobs and tables, data paging, and data analysis

Securing Multi-tenant Applications
Protecting sensitive data, protecting session tokens, authentication and authorization

Managing and Monitoring Multi-tenant Applications
ALM, endpoint protection, provisioning new tenants, customization, billing

"The Tailspin Scenario" introduces you to the Tailspin company and the Surveys application. It provides an architectural overview of the Surveys application; the following chapters provide more information about how Tailspin designed and implemented the Surveys application for the cloud. Reading this chapter will help you understand Tailspin's business model, its strategy for adopting the cloud platform, and some of its concerns. It will also help you to understand some of the fundamental choices Tailspin had to make when designing the application.

"Hosting a Multi-tenant Application on Windows Azure" discusses the major considerations that surround architecting and building multi-tenant applications to run on Windows Azure. It describes the benefits of a multi-tenant architecture and the trade-offs that you must consider. This chapter provides a conceptual framework that helps you understand the topics that are discussed in more detail in the subsequent chapters.

"Choosing a Multi-tenant Data Architecture" describes the important factors to consider when designing the data model for multi-tenant applications. The major factors are how you can partition data, plan for extensibility and scalability, and how you can apply your design using Windows Azure storage and a relational database. The chapter describes how the Surveys application stores data in both Windows Azure tables and blobs, and how the developers at Tailspin designed their storage classes to be extensible and testable. It also describes the role that Windows Azure SQL Database plays in the Surveys application.

"Partitioning Multi-tenant Applications" describes how you can partition your application code for multiple tenants. This includes how you can use Cloud Services web and worker roles, queues, and the Model View Controller pattern to best effect in a multi-tenant application. The chapter also discusses issues around caching, and how Tailspin solved some specific problems related to implementing session state.

"Maximizing Availability, Scalability, and Elasticity" describes techniques you can use to get the best performance and responsiveness for your applications, especially when they are designed to support multiple tenants. The chapter covers topics such as hosting the application in multiple geographic locations, using the Content Delivery Network (CDN) to cache content, read and write patterns using queues, paging and displaying data, and autoscaling the role instances.

"Securing Multi-tenant Applications" describes authentication and authorization scenarios for multi-tenant applications when supporting individual subscribers and users, and through trust relationships. It also examines how Tailspin implemented protection and isolation of sensitive data, and how it protects session tokens.

"Managing and Monitoring Multi-tenant Applications" examines application lifecycle management (ALM) considerations for multi-tenant applications, how Tailspin manages and monitors the application, and how the application supports on-boarding, customization, and billing for customers.

WHAT YOU NEED TO USE THE CODE

These are the system requirements for running the scenarios:

- Microsoft Windows 7 with Service Pack 1, Microsoft Windows 8, Microsoft Windows Server 2008 R2 with Service Pack 1, or Microsoft Windows Server 2012 (32-bit or 64-bit editions).
- *Microsoft .NET Framework version 4.0.*

- *Microsoft Visual Studio* 2010 Ultimate, Premium, or Professional edition with Service Pack 1 installed, or Visual Studio 2012 Ultimate, Premium, or Professional edition.
- *Windows Azure SDK* (includes the Windows Azure Tools for Visual Studio). See the Release Notes for information on the specific version required.
- *Microsoft SQL Server 2012, SQL Server Express 2012, SQL Server 2008, or SQL Server Express 2008.* See the Release Notes for information on specific versions depending on your operating system.
- *ASP.NET MVC 4 Framework.*
- *Windows Identity Foundation.* This is required for claims-based authorization.
- *WebAii testing framework.* This is required only if you want to run the functional tests. Place the assembly **ArtOfTest.WebAii.dll** in the **Lib\WebAii** folder of the examples.

Other components and frameworks required by the examples are installed using NuGet when you run the solutions. See the Release Notes included with the examples for instructions on installing and configuring them.

Where to Go for More Information

There are a number of resources listed in text throughout the book. These resources will provide additional background, bring you up to speed on various technologies, and so forth. For your convenience, there is a bibliography online that contains all the links so that these resources are just a click away.

You can find the bibliography at: *http://msdn.microsoft.com/library/jj871057.aspx.*

Who's Who

A panel of experts comments on Tailspin's development efforts and on the example application provided for this guide. The panel includes a cloud specialist, a software architect, a software developer, and an IT professional. The delivery of the application can be considered from each of these points of view. The following table lists these experts.

Bharath is a cloud specialist. He checks that a cloud-based solution will work for a company and provide tangible benefits. He is a cautious person, for good reasons.

"Implementing a single-tenant application for the cloud is easy. Realizing the benefits that a cloud-based solution can offer to multi-tenant applications is not always so straight-forward."

Jana is a software architect. She plans the overall structure of an application. Her perspective is both practical and strategic. In other words, she considers the technical approaches that are needed today and the direction a company needs to consider for the future."

"It's not easy to balance the needs of the company, the users, the IT organization, the developers, and the technical platforms we rely on."

Markus is a senior software developer. He is analytical, detail-oriented, and methodical. He's focused on the task at hand, which is building a great cloud-based application. He knows that he's the person who's ultimately responsible for the code.

"For the most part, a lot of what we know about software development can be applied to the cloud. But, there are always special considerations that are very important."

Poe is an IT professional who's an expert in deploying and running applications in the cloud. Poe has a keen interest in practical solutions; after all, he's the one who gets paged at three o'clock in the morning when there's a problem.

"Running applications in the cloud that are accessed by thousands of users involves some big challenges. I want to make sure our cloud apps perform well, are reliable, and are secure. The reputation of Tailspin depends on how users perceive the applications running in the cloud."

If you have a particular area of interest, look for notes provided by the specialists whose interests align with yours.

Acknowledgments

On March 4, 2010 I saw an email from our CEO, Steve Ballmer, in my inbox. I don't normally receive much email from him, so I gave it my full attention. The subject line of the email was: "We are all in," and it summarized the commitment of Microsoft to cloud computing. If I needed another confirmation of what I already knew, that Microsoft is serious about the cloud, there it was.

My first contact with what eventually became Windows Azure, and other components of what is now called the Windows Azure platform, was several years ago. I was in the Developer & Platform Evangelism (DPE) team, and my job was to explore the world of software delivered as a service. Some of you might even remember a very early mockup I developed in late 2007, called Northwind Hosting. It demonstrated many of the capabilities that the Windows Azure platform offers today. (Watching an initiative I've been involved with since the early days become a reality makes me very, very happy.)

In February 2009, I left DPE and joined the patterns & practices team. My mission was to lead the "cloud program" - a collection of projects that examined the design challenges of building applications for the cloud. When the Windows Azure platform was announced, demand for guidance about it skyrocketed.

As we examined different application development scenarios, it became quite clear that identity management is something you must get right before you can consider anything else. It's especially important if you are a company with a large portfolio of on-premises investments, and you want to move some of those assets to the cloud. This describes many of our customers.

In December 2009, we released the first edition of A Guide to Claims-Based Identity and Access Control. This was patterns & practices's first deliverable, and an important milestone in our cloud program. We followed it with Moving Applications to the Cloud. This was the first in a three part series of guides that address development in Windows Azure. Both of these guides have been regularly updated as Windows Azure evolves.

Windows Azure is special in many ways. One is the rate of innovation. The various teams that deliver all of the platform's systems proved that they could rapidly ship new functionality. To keep up with them, I felt we had to develop content very quickly. We decided to run our projects in two-months sprints, each one focused on a specific set of considerations.

This guide covers a Greenfield scenario: designing and developing new multi-tenant applications for the Windows Azure platform. This follows on from the previous guide that focused on how to move an existing application to the Windows Azure platform. As in the previous guides, we've developed a fictitious case study that explains, step by step, the challenges our customers are likely to encounter.

I want to start by thanking the following subject matter experts and contributors to this guide: Dominic Betts (Content Master Ltd), Scott Densmore (Microsoft Corporation), Ryan Dunn, Steve Marx, and Matias Woloski. Dominic has the unusual skill of knowing a subject in great detail and of finding a way to explain it to the rest of us that is precise, complete, and yet simple to understand. Scott brought us a wealth of knowledge about how to build scalable Windows Azure applications, which is what he did before he joined my team. He also brings years of experience about how to build frameworks and tools for developers. I've had the privilege of working with Ryan in previous projects, and I've always benefited from his acuity, insights, and experience. As a Windows Azure evangelist, he's been able to show us what customers with very real requirements need. Steve is a technical strategist for Windows Azure. He's been instrumental in shaping this guide. We rely on him to show us not just what the platform can do today but how it will evolve. This is important because we want to provide guidance today that is aligned with longer-term goals. Last but not least, Matias is a veteran of many projects with me. He's been involved with Windows Azure since the very first day, and his efforts have been invaluable in creating this guide.

As it happens with all our written content, we have sample code for most of the chapters. They demonstrate what we talk about in the guide. Many thanks to the project's development and test teams for providing a good balance of technically sound, focused and simple-to-understand code: Masashi Narumoto (Microsoft Corporation), Scott Densmore (Microsoft Corporation), Federico Boerr (Southworks), Adrián Menegatti (Southworks), Hanz Zhang (Microsoft Corporation), Ravindra Mahendravarman (Infosys Ltd.), Rathi Velusamy (Infosys Ltd.).

Our guides must not only be technically accurate but also entertaining and interesting to read. This is no simple task, and I want to thank Dominic Betts (Content Master Ltd), RoAnn Corbisier (Microsoft Corporation), Alex Homer (Microsoft Corporation), and Tina Burden from the writing and editing team for excelling at this.

The visual design concept used for this guide was originally developed by Roberta Leibovitz and Colin Campbell (Modeled Computation LLC) for A Guide to Claims-Based Identity and Access Control. Based on the excellent responses we received, we decided to reuse it for this guide. The guide design was created by John Hubbard (eson). The cartoon faces were drawn by the award-winning Seattle-based cartoonist Ellen Forney. The technical illustrations were adapted from my Tablet PC mockups by Rob Nance and Katie Niemer.

All of our guides are reviewed, commented upon, scrutinized, and criticized by a large number of customers, partners, and colleagues. We also received feedback from the larger community through our CodePlex website. The Windows Azure platform is broad and spans many disciplines. We were very fortunate to have the intellectual power of a very diverse and skillful group of readers available to us.

I also want to thank all of these people who volunteered their time and expertise on our early content and drafts. Among them, I want to mention the exceptional contributions of David Aiken (Microsoft Corporation), Graham Astor (Avanade), Edward Bakker (Inter Access), Vivek Bhatnagar (Microsoft Corporation), Patrick Butler Monterde (Microsoft Corporation), Shy Cohen, James Conard (Microsoft

Corporation), Brian Davis (Longscale), Aashish Dhamdhere (Windows Azure, Microsoft Corporation), Andreas Erben (DAENET), Giles Frith, Eric L. Golpe (Microsoft Corporation), Johnny Halife (Southworks), Simon Ince (Microsoft Corporation), Joshy Joseph (Microsoft Corporation), Andrew Kimball, Milinda Kotelawele (Longscale), Mark Kottke (Microsoft Corporation), Chris Lowndes (Avanade), Dianne O'Brien (Windows Azure, Microsoft Corporation), Steffen Vorein (Avanade), Michael Wood (Strategic Data Systems).

I hope you find this guide useful!

Eugenio Pace

Senior Program Manager – *patterns & practices*

Microsoft Corporation

ACKNOWLEDGEMENTS OF CONTRIBUTORS TO THE THIRD EDITION

Windows Azure is an evolving platform. We originally published the first edition of this guide in 2010, demonstrating a basic set of Windows Azure features. I'm now pleased to release the third edition of this guide, which is more tailored to multi-tenant scenario. This new edition describes common challenges in the multi-tenant Software as a Service applications such as partitioning data, data extensibility, automated provisioning, customizing to multiple tenants, and so on.

As our scope increased, we also added new community members and industry experts who have provided significant help throughout the development of this edition. I want to acknowledge the exceptional contributions of following people: Dominic Betts (ContentMaster), Alex Homer (Microsoft Corporation), Alejandro Jezierski (Southworks), Mauro Krikorian (Southworks), Jorge Rowies (Southworks), Marcos Castany (Southworks), Hanz Zhang (Microsoft Corporation), Rathi Velusamy (Infosys), RoAnn Corbisier (Microsoft Corporation), Nelly Delgado (Microsoft Corporation), Eugenio Pace (Microsoft Corporation), Carlos Farre (Microsoft Corporation), Trent Swanson (Full Scale 180 Inc.), Ercenk Keresteci (Full Scale 180 Inc.), Jane Sinyagina (Microsoft Corporation), Hatay Tuna (Microsoft Corporation), Patrick Butler Monterde (Microsoft Corporation), and Michael Wood. I also want to thank everyone who participated in our CodePlex community site.

Masashi Narumoto

Senior Program Manager – *patterns & practices*

Microsoft Corporation

Redmond, October 2012

1   The Tailspin Scenario

This chapter introduces a fictitious company named Tailspin. It describes Tailspin's plans to launch a new online service named Surveys that will enable other companies or individuals to conduct their own online surveys. The chapter also describes why Tailspin wants to host its survey application on Windows Azure. As with any company considering this process, there are many issues to consider and challenges to be met, particularly because this is the first time Tailspin is using the cloud. The chapters that follow this one show how Tailspin architected and built its survey application to run on Windows Azure.

The Tailspin Company

Tailspin is a startup ISV company of approximately 20 employees that specializes in developing solutions using Microsoft technologies. The developers at Tailspin are knowledgeable about various Microsoft products and technologies, including the .NET Framework, ASP.NET MVC, SQL Server, and Visual Studio. These developers are aware of Windows Azure but have not yet developed any complete applications for the platform.

The Surveys application is the first of several innovative online services that Tailspin wants to take to market. As a startup, Tailspin wants to develop and launch these services with a minimal investment in hardware and IT personnel. Tailspin hopes that some of these services will grow rapidly, and the company wants to have the ability to respond quickly to increasing demand. Similarly, it fully expects some of these services to fail, and it does not want to be left with redundant hardware on its hands.

Tailspin's Strategy

Tailspin is an innovative and agile organization, well placed to exploit new technologies and the business opportunities offered by the cloud. As a startup, Tailspin is willing to take risks and use new technologies when it implements applications. Tailspin's plan is to embrace the cloud and gain a competitive advantage as an early adopter. It hopes to rapidly gain some experience, and then quickly expand on what it has learned. This strategy can be described as "try, fail fast, learn, and then try again." Tailspin has decided to start with the Surveys application as its first cloud-based service offering.

The Surveys Application

The Surveys application enables Tailspin's customers to design a survey, publish the survey, and collect the results of the survey for analysis. A survey is a collection of questions, each of which can be one of several types such as multiple-choice, numeric range, or free text. Customers begin by creating a subscription with the Surveys service, which they use to manage their surveys and to apply branding by using styles and logo images.

Customers can also select a geographic region for their account, so that they can host their surveys as close as possible to the survey audience. In addition, Tailspin enables premium customers to add custom fields to surveys for integration with the customers' own systems. The Surveys application allows users to try out the application for free, and to sign up for one of several different packages that offer different collections of services for a monthly fee.

Figure 1 illustrates the Surveys application and highlights the three different groups of users who interact with application. All three websites interact with the core services that comprise the Surveys application and provide access to the application's data storage.

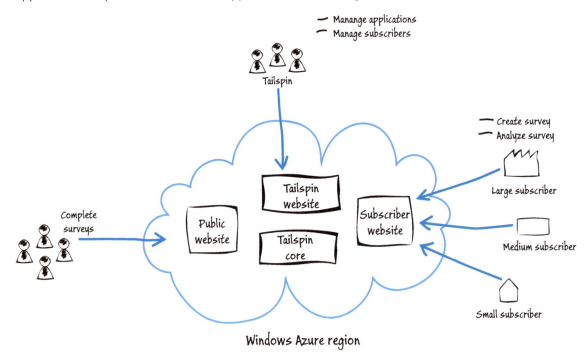

FIGURE 1
The Surveys application

Customers who sign up and become subscribers to the Surveys service (or who are using a free trial) access the Subscriber website that enables them to design their own surveys, apply branding and customization, and collect and analyze the survey results. Depending on the package they select, they have access to different levels of functionality within the Surveys application. Tailspin expects its subscribers to be of various sizes and from all over the world; and they can select a geographic region for their account and surveys.

Tailspin wants to design the service in such a way that most of the administrative and configuration tasks are "self-service" and performed by the subscriber with minimal intervention by Tailspin staff.

In the world of Software as a Service (SaaS), subscribers are commonly known as "tenants." We commonly refer to applications like Tailspin Surveys as "multi-tenant" applications. When we talk about Tailspin's "customers" we are referring to the subscribers or tenants, and we use this terminology throughout most of this guide.

The public website enables the people participating in the survey to complete their responses to the survey questions. The survey creator will inform their survey audience of the URL to visit to complete the survey.

The Tailspin website enables staff at Tailspin to manage the application and manage the subscriber accounts. Note that this website is not included in the example application you will see discussed in this guide, which focuses on the public and the subscriber website functionality.

> For information about building a Windows Phone 7 client application for the Tailspin Surveys application, see **"Developing an Advanced Windows Phone 7.5 App that Connects to the Cloud."**

Tailspin's Goals and Concerns

Tailspin faces several challenges, both as an organization and with the Surveys application in particular. First, subscribers might want to create surveys associated with a product launch or a marketing campaign, or the surveys might be seasonal—perhaps associated with a holiday period. Often, subscribers who use the Surveys application will want to set up these surveys with a very short lead-time. Surveys will usually run for a fixed, short period of time but may have a large number of respondents.

This means that usage of the Surveys application will tend to spike and Tailspin will have very little warning of when these spikes will occur. Tailspin wants to be able to offer the Surveys application to subscribers around the world, and because of the nature of the Surveys application with sudden spikes in demand, it wants to be able to quickly expand or contract its infrastructure in different geographical locations. It doesn't want to purchase and manage its own hardware, or maintain sufficient capacity to meet peak demand. Neither does Tailspin want to sign long-term contracts with hosting providers for capacity that it will use for only part of the time.

Resource elasticity and geo-distribution are key properties of Windows Azure.

Tailspin wants to be able to maintain its competitive advantage by rapidly rolling out new features for existing services, or gain competitive advantage by being first to market with new products and services.

With the Surveys application, Tailspin wants to offer its subscribers a reliable, customizable, and flexible service for creating and conducting online surveys. It must provide its subscribers with the ability to create surveys using a range of question types, and the ability to brand the surveys using corporate logos and color schemes.

Tailspin wants to be able to offer different packages (at different prices) to subscribers, based on each subscriber's specific requirements. Tailspin wants to offer its larger subscribers the ability to integrate the Surveys application into that subscriber's own infrastructure. For example, integration with the subscriber's own identity infrastructure could provide single sign-on (SSO), or enable multiple users to manage surveys or access billing information. Integration with the subscriber's own business intelligence (BI) systems could provide for a more sophisticated analysis of survey results. For small subscribers who don't need, or can't use, the sophisticated integration features, a basic package might include an authentication system. The range of available packages should also include a free trial to enable subscribers to try the Surveys application before they purchase a subscription.

The subscriber and public websites also have different scalability requirements. It is likely that thousands of users might complete a survey, but only a handful of users from each subscriber will edit existing surveys or create new surveys. Tailspin wants to optimize the resources for each of these scenarios.

The Tailspin business model is to charge subscribers a monthly fee for a service such as the Surveys application and, because of the global market they are operating in, Tailspin wants its prices to be competitive. Tailspin must then pay the actual costs of running the application, so in order to maintain its profit margin Tailspin must tightly control the running costs of the services it offers to subscribers.

> *In this scenario, Tailspin's customers (the subscribers) are **not** Windows Azure customers. Subscribers pay Tailspin, who in turn pays Microsoft for the subscribers' use of Windows Azure services.*

Tailspin wants to ensure that subscribers' data is kept safe. For example, a subscriber's data must be private to that subscriber, there must be multiple physical copies of the survey data, and subscribers should not be able to lose data by accidently deleting a survey. In addition, all existing survey data must be preserved whenever Tailspin updates the application.

Finally, Tailspin would like to be able to leverage the existing skills of its developers to build the Surveys application, and minimize any necessary retraining.

THE SURVEYS APPLICATION ARCHITECTURE

To achieve the goals of the Surveys application, Tailspin decided to implement the application as a cloud-based service using Windows Azure. Figure 2 shows a high-level view of this architecture.

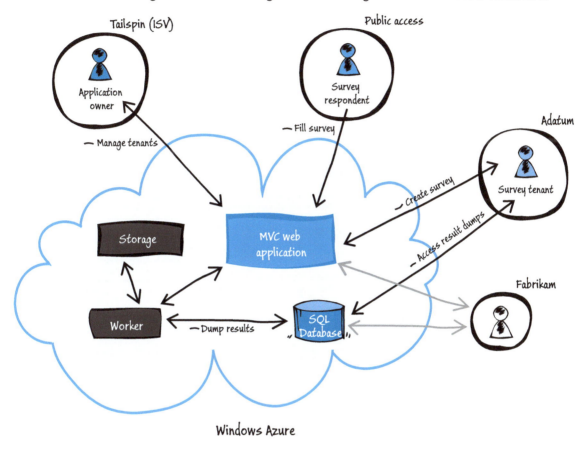

FIGURE 2
The Surveys application architecture

The architecture of the Surveys application is straightforward, and one that many other Windows Azure applications use. The core of the application uses Windows Azure web roles, worker roles, and storage. Figure 2 shows the three groups of users who access the application: the application owner, the public, and the subscribers to the Surveys service (in this example, the tenants Adatum and Fabrikam). It also highlights how the application uses Windows Azure SQL Database to provide a mechanism for subscribers to dump their survey results into a relational database so that they can analyze the results in detail.

This guide discusses how Tailspin designed and implemented the Surveys application as a multi-tenant application. It addresses common multi-tenant challenges such as partitioning, extensibility, provisioning, testability, and customization. For example, the guide describes how Tailspin handles the integration of the application's authentication mechanism with a subscriber's own security infrastructure by using a "federated identity with multiple partners" model. The guide also covers the reasoning behind the decision to use a hybrid data model that comprises both Windows Azure storage and Windows Azure SQL Database.

Other topics covered in this guide include how the application uses Windows Azure Caching to ensure the responsiveness of the public website for survey respondents, how the application automates the on-boarding and provisioning process, how the application leverages the Windows Azure geographic location feature, and the subscriber billing model that Tailspin adopted for the Surveys application.

Tailspin will build the application using Visual Studio, ASP.NET MVC, and the .NET Framework. The following table will help you to identify the areas of the guide that correspond to the various features of the application and the Windows Azure services it uses.

Chapter	Topic Areas	Relevant Technologies
2 – "Hosting a Multi-Tenant Application on Windows Azure"	Choosing a single or multi-tenant architecture. Considerations for stability, scalability, authentication and authorization, ALM, SLAs, monitoring, code partitioning, billing, and customization.	
3 – "Choosing a Multi-Tenant Data Architecture"	Considerations for Windows Azure storage, SQL Server, and SQL Database. Using SQL Federations. Data partitioning strategies. Data architecture, extensibility, and scalability. Displaying data in the UI.	Windows Azure storage tables and blobs. Microsoft SQL Server. Windows Azure SQL Database.
4 – "Partitioning Multi-Tenant Applications"	Partitioning queues and worker roles. Prioritizing some tenants. Accessing the web roles as a tenant. Session management.	Windows Azure web and worker roles. Windows Azure storage queues. ASP.NET MVC.
5 – "Maximizing Availability, Scalability, and Elasticity"	Geo-location and routing. The delayed write pattern. Background processes. Caching static data. Auto scaling role instances.	Windows Azure worker roles. Windows Azure storage queues. Windows Azure Caching. Windows Azure Traffic Manager. Enterprise Library Integration Pack for Windows Azure.
6 – "Securing Multi-Tenant Applications"	Authentication and authorization strategies. Protecting sensitive data. Protecting session tokens.	Windows Identity Framework. Claims-based authentication and authorization. Windows Azure Active Directory.
7 – "Managing and Monitoring Multi-Tenant Applications"	Application lifecycle management, including testing, monitoring, and managing the application. Automated provisioning and trial subscriptions. Per tenant customization. Billing subscribers.	Windows Azure diagnostics. Windows Azure PowerShell Cmdlets. Windows Azure Endpoint Protection.

MORE INFORMATION

All links in this book are accessible from the book's online bibliography available at:
http://msdn.microsoft.com/library/jj871057.aspx.

Overview of *Windows Azure features.*

Data Storage Offerings on the Windows Azure Platform.

Introducing Windows Azure provides a list of features and services.

For information about building a Windows Phone 7 client application for the Tailspin Surveys application, see the guide *"Developing an Advanced Windows Phone 7.5 App that Connects to the Cloud."*

The guide *"Moving Applications to the Cloud"* explores techniques for migrating existing applications to Windows Azure.

The guide *"Building Hybrid Applications in the Cloud"* describes the scenarios for and usage of many Windows Azure features.

2   Hosting a Multi-Tenant Application on Windows Azure

This chapter discusses some of the issues that surround architecting and building multi-tenant applications to run on Windows Azure. A highly scalable, cloud-based platform offers a compelling set of features for building services that many users will pay a subscription to use. A multi-tenant architecture where multiple users share the application enables economies of scale as users share resources, but at the cost of a more complex application that has to manage multiple users independently of each other.

This chapter does not focus specifically on Tailspin or the Surveys application, but it uses the scenario described in the previous chapter to illustrate some of the factors that you might consider when choosing how to implement a multi-tenant application on Windows Azure.

This chapter provides a conceptual framework that helps you understand some of the topics discussed in more detail in the subsequent chapters of this guide.

Goals and Requirements

This section outlines some of the goals and requirements that are common to many multi-tenant applications. Some may not be relevant in some specific scenarios, and the importance of individual goals and requirements will differ in each scenario. For example, not all multi-tenant applications require the same level of customizability by the tenant or face the same regulatory constraints.

It is also useful to consider the goals and requirements for a multi-tenant application from the perspective of both the tenant and the provider.

The Tenant's Perspective

Multiple tenants share the use of a multi-tenant application, but different tenants may have different goals and requirements. A tenant is unlikely to be interested how the provider implements the multi-tenancy, but will expect the application to behave as if the tenant is its sole user. The following provides a list of the most significant goals and requirements from a tenant's perspective.

- **Isolation.** This is the most important requirement in a multi-tenant application. Individual tenants do not want the activities of other tenants to affect their use of the application. They also need to be sure that other tenants cannot access their data. Tenants want the application to appear as though they have exclusive use of it.

- **Availability.** Individual tenants want the application to be constantly available, perhaps with guarantees defined in an SLA. Again, the activities of other tenants should not affect the availability of the application.
- **Scalability.** Even though multiple tenants share a multi-tenant application, an individual tenant will expect the application to be scalable and be able to meet his level of demand. The presence and actions of other tenants should not affect the performance of the application.
- **Costs.** One of the expectations of using a multi-tenant application is that the costs will be lower than running a dedicated, single-tenant application because multi-tenancy enables the sharing of resources. Tenants also need to understand the charging model so that they can anticipate the likely costs of using the application.
- **Customizability.** An individual tenant may require the ability to customize the application in various ways such as adding or removing features, changing colors and logos, or even adding their own code or script.
- **Regulatory Compliance.** A tenant may need to ensure that the application complies with specific industry or regulatory laws and limitations, such as those that relate to storing personally identifiable information (PII) or processing data outside of a defined geographical area. Different tenants may have different requirements.

The Provider's Perspective

The provider of the multi-tenant application will also have goals and requirements. The following provides a list of the most significant goals and requirements from a provider's perspective.

- **Meeting the tenants' goals and requirements.** The provider must ensure that the application meets the tenants' expectations. A provider may offer a formal SLA that defines how the application will meet the tenants' requirements.
- **Profitability.** If the provider offers the application as a commercial service, the provider will want to obtain an appropriate rate of return on the investment it has made in developing and providing the service. Revenue from the application must be sufficient to cover both the capital and running costs of the application.
- **Billing.** The provider needs a way to bill the tenants. This may require the application to monitor resource usage if the provider does not want to use a fixed rate charging approach. An example of a fixed rate approach would be if Tailspin charges each tenant a monthly fee for using the Surveys application. An alternative is that Tailspin charges each tenant based on the number of survey responses it collects, or on some other usage metric.

- **Multiple service levels.** The provider may want to offer different versions of a service at different monthly rates, such as a standard or a premium subscription. These different subscription levels may include different functions, different usage limitations, have different SLAs, or specify some combination of these factors.

- **Provisioning.** The provider must be able to provision new tenants for the application. If there are a small number of tenants, this may be a manual process. For multi-tenant applications with a large number of tenants, it is usually necessary to automate this process by enabling self-service provisioning.

- **Maintainability.** The provider must be able to upgrade the application and perform other maintenance tasks while multiple tenants are using it.

- **Monitoring.** The provider must be able to monitor the application at all times to identify any problems and to troubleshoot them. This includes monitoring how each tenant is using the application.

- **Automation.** In addition to automated provisioning, the provider may want to automate other tasks in order to provide the required level of service. For example, the provider may want to automate the scaling of the application by dynamically adding or removing resources as and when they are required.

Single Tenant vs. Multiple Tenant

One of the first architectural decisions that the team at Tailspin had to make about how the Surveys application could best support multiple subscribers was whether it should be a single-tenant or multi-tenant application. Figure 1 shows the difference between these approaches at a high-level. The single-tenant model has a separate physical instance of the application for each subscriber, while the multi-tenant model has a single physical instance of the application shared by many subscribers.

It's important to note that the multi-tenant model still offers separate views of the application's data to its users. In the Surveys application, Client B must not be able to see or modify Client A's surveys or data. Tailspin, as the owner of the application, will have full access to all the data stored in the application.

Multi-instance, single tenant

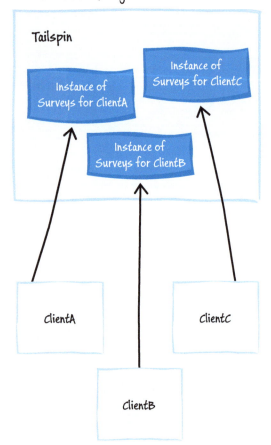

Single instance, multi-tenant

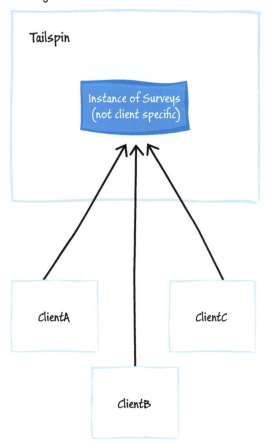

FIGURE 1
Logical view of single tenant and multiple tenant architectures

This diagram shows logical instances of the Surveys application. In practice, you can implement each logical instance as multiple physical instances to scale the application.

MULTI-TENANCY ARCHITECTURE IN WINDOWS AZURE

In Windows Azure, the distinction between the multi-tenant model and the single-tenant model is not as straightforward as that shown in Figure 1 because an application in Windows Azure can consist of many elements, each of which can be single tenanted or multiple tenanted. For example, if an application has a user interface (UI) element, a services element, and a storage element, a possible design could look like that shown in Figure 2.

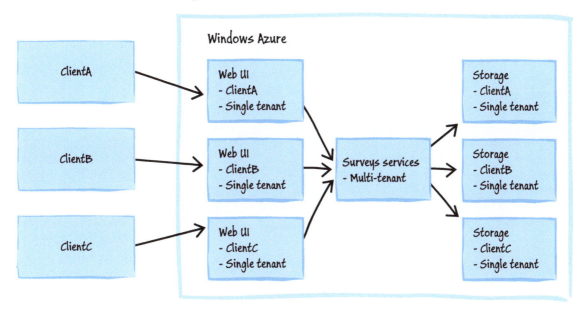

FIGURE 2
Sample architecture for Windows Azure

This is not the only possible design, but it illustrates that you don't have to make the same choice of either a single-tenancy or a multi-tenancy model for every element in your application. In practice, a Windows Azure application consists of many more elements than shown in Figure 2 such as queues, caches, and virtual networks that might have a single-tenant or a multi-tenant architecture.

> *Chapter 3, "Choosing a Multi-Tenant Data Architecture," looks at the issues that relate to data storage and multi-tenancy. Chapter 4, "Partitioning Multi-Tenant Applications," looks at the issues that relate to partitioning Windows Azure roles, caches, and queues. Chapter 6, "Securing Multi-Tenant Applications," and Chapter 7, "Managing and Monitoring Multi-Tenant Applications," cover multi-tenancy in other application elements.*

Should you design your Windows Azure application to be single-tenant or multi-tenant? There's no right or wrong answer but, as you will see in the following section, there are a number of factors that can influence your choice.

SELECTING A SINGLE-TENANT OR MULTI-TENANT ARCHITECTURE

This section introduces some of the criteria that an architect would consider when deciding on a single-tenant or multi-tenant design. The guide revisits many of these topics in more detail, and with specific reference to Tailspin and the Surveys application, in later chapters. The relative importance of the different criteria will vary for different application scenarios.

This chapter focuses on application architecture, management, and financial considerations. Chapter 3, "Choosing a Multi-Tenant Data Architecture," explores the topics you must consider when choosing a suitable data architecture for a multi-tenant application.

Architectural Considerations

The architectural requirements of your application will influence your choice of a single-tenant or multi-tenant architecture.

The focus of this guide is on building a multi-tenant application using Windows Azure cloud services: web and worker roles. However, the architectural considerations addressed in this chapter, and many of the design decisions that Tailspin faced during the implementation of the Surveys application discussed in subsequent chapters, are equally relevant to other hosting choices for your multi-tenant application. For example, if you decide to build your multi-tenant application using Windows Azure Web Sites or to deploy it to Windows Azure Virtual Machines you will face many of the same challenges that Tailspin faced building the Surveys application for deployment to Windows Azure Cloud Services.

For a detailed discussion of the Infrastructure as a Service (IaaS) approach offered by Windows Azure Virtual Machines, you should read Chapter 2, *"Getting to the Cloud,"* in the guide *"Moving Applications to the Cloud."* Chapter 3, *"Moving to Windows Azure Cloud Services,"* in that guide discusses using Windows Azure Web Sites to host your application in the cloud.

Application Stability

A multi-tenant application is more vulnerable to instance failure than a single-tenant application. If a single-tenant instance fails, only the user of that instance is affected. If the multi-tenant instance fails, all users are affected. However, Windows Azure can help to mitigate this risk by enabling you to deploy multiple, identical instances of the Windows Azure roles that make up your application (this is really a multi-tenant, multi-instance model).

Windows Azure load balances requests across those role instances, and you must design your application so that it functions correctly when you deploy multiple instances. For example, if your application uses session state you must make sure that each web role instance can access the state for any user. In addition, the tasks that a worker role performs must function correctly when Windows Azure can select any instance of the role to handle a particular task. Windows Azure monitors your role instances and automatically restarts any that have failed.

For the Windows Azure SLA to apply to your application, you must have at least two instances of each role type running. For more information, see "Service Level Agreements."

Windows Azure can throttle access to resources, making them temporarily unavailable. Typically, this happens when there is high contention for a resource. Your Windows Azure application should detect when it is being throttled, and take appropriate action such as retrying the operation after a short delay.

Making the Application Scalable

The scalability of an application running on Windows Azure depends largely on being able to deploy multiple instances of your web and worker roles, while being able to access the same data from those instances. Both single-tenant and multi-tenant applications use this feature to scale out when they run on Windows Azure. Windows Azure also offers various instance sizes that enable you to scale up or scale down individual instances.

The *Transient Fault Handling Application Block*, available as a separately installable part of the Enterprise Library 5.0 Integration Pack for Windows Azure, can handle in a standard and configurable way the transient faults that may occur because of throttling.

Figure 3 shows how you can scale out the application by running a variable number of instances. In Windows Azure cloud services, these would be multiple instances of your web and worker roles.

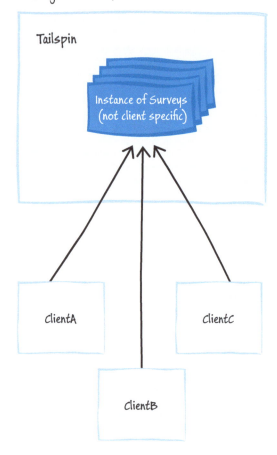

Multi-instance, single tenant

Tailspin

Instance of Surveys for ClientA

Instance of Surveys for ClientC

Instance of Surveys for ClientB

ClientA

ClientC

ClientB

Single instance, multi-tenant

Tailspin

Instance of Surveys (not client specific)

ClientA

ClientC

ClientB

FIGURE 3
Scaling out a multi-tenant application

In Windows Azure, the preferred way to adapt your application to manage varying load is to scale out by adding additional nodes, rather than scale up by using larger nodes. This enables you to add or remove capacity as and when it's needed without interruption to services. You can use frameworks or scripts to automatically add and remove instances based on a schedule, or in response to changes in demand. The *Autoscaling Application Block*, available as part of the Enterprise Library 5.0 Integration Pack for Windows Azure, is an example of such a framework.

For some applications, you may not want to have all your subscribers sharing just one multi-tenant instance. For example, you may want to group your subscribers based on the functionality they use or their expected usage patterns, and then optimize each instance for the subscribers who are using it. In this case, you may need to have two or more copies of your multi-tenanted application deployed in different cloud services or Windows Azure accounts.

Figure 4 illustrates a scenario where premium subscribers share one instance of the application, and standard subscribers share another instance. Note that you can scale each instance independently of the other.

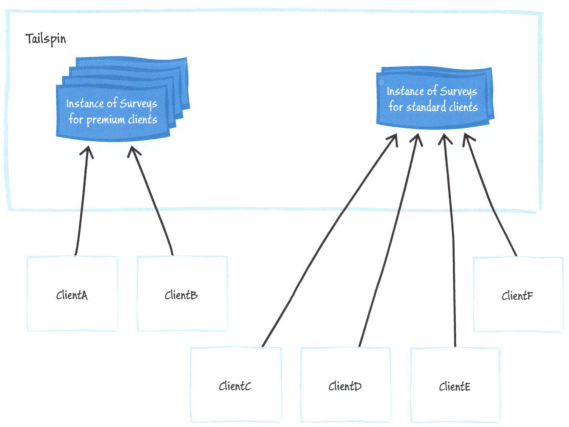

FIGURE 4
Using multiple multi-tenant instances

Although the model shown in Figure 4 makes it easy to scale the application for premium subscribers independently of standard subscribers, it is not the only way to handle different subscription levels. For example, if both premium and standard subscribers shared the same instance you could implement an algorithm that gives preference to premium users, ensuring that their workload and tasks are given priority within the instance. By providing configuration parameters, you could adjust the algorithm dynamically.

If you use an autoscaling solution when your application has multiple tenants, you need to consider any limits that you want to place on the scalability of your application because each running role instance will accrue charges. It's possible that the activities of a tenant could cause a large number of instances to automatically start. With fixed rate charging, this could result in high costs for the provider. With usage based charging, this could result in high costs for the tenant.

You may want to consider using Windows Azure Caching and Windows Azure Traffic Manager to enhance the scalability of your application. In addition to providing output caching and data caching, Windows Azure Caching includes a highly scalable session provider for use in ASP.NET applications. Traffic Manager enables you to control the distribution of traffic to multiple Windows Azure deployments, even if those deployments are running in different data centers.

Chapter 4, "Partitioning Multi-Tenant Applications," of this guide contains more information about how you can use Windows Azure Caching. Chapter 5, "Maximizing Availability, Scalability, and Elasticity," of this guide discusses scalability and related topics, including using Windows Azure Traffic Manager and how you can automatically scale instances of your application using the Enterprise Library Autoscaling Application Block.

Resource Limitations and Throttling

Individual elements of your application architecture will have specific limitations, such as the maximum throughput of the message queuing element (Windows Azure storage queues or Windows Azure Service Bus), or the maximum number of transactions per second supported by the data storage system used in your application. These resource limitations may place constraints on the number of tenants who can share a particular instance. You must understand the resource limitations and quotas in relation to the likely usage patterns of your tenants so that these resource limitations do not affect overall performance of the application.

Remember that Windows Azure is itself a multi-tenant service, and one of the ways that it manages contention for resources by its tenants is to use throttling.

> *Some of the quotas associated with Windows Azure Service Bus include the queue/topic size, the number of concurrent connections, and the number of topics/queues per service namespace.*

Furthermore, many resources in the cloud, such as message queues and storage systems, may throttle usage at certain times when they are under high load or encounter spikes of high activity. You should try to design your application so that it is unlikely to be throttled, but it must still be resilient if it does encounter throttling.

Geo-location

If your application has tenants from multiple geographic locations, giving them access to resources in their country or region can help to improve performance and reduce latency. In this scenario, you should consider a partitioning scheme that uses location to associate tenants with specific resource. In Windows Azure, whenever you create a resource such as a storage account, a cloud service, or a service namespace you can specify the geographic location where the resource will be hosted.

Service Level Agreements

You may want to offer a different Service Level Agreement (SLA) with the different subscription levels for the service. If subscribers with different SLAs are sharing the same multi-tenant instance, you should aim to meet the highest SLA, thereby ensuring that you also satisfy the lower SLAs for other subscribers.

However, if you have a limited number of different SLAs, you could put all the subscribers that share the same SLA into the same multi-tenant instance and make sure that the instance has sufficient resources to satisfy the requirements of the SLA.

The Legal and Regulatory Environment

For some applications, you may need to take into account specific regulatory or legal issues. This may require some differences in functionality, specific legal messages to be displayed in the UI, guaranteed separate databases for storage, or storage located in a specific county or region. This may again lead to having separate multi-tenant deployments for groups of subscribers, or it may even require a single-tenant architecture.

Handling Authentication and Authorization

You may want to provide your own authentication and authorization systems for your cloud application that require subscribers to set up accounts for the users who will interact with the application. However, subscribers may prefer to use an identity they have already established with an existing authentication system (such as a Microsoft or a Google account, or an account in their own Active Directory) and avoid having to create a new set of credentials for your application.

In a multi-tenant application, this implies being able to support multiple authentication providers, and it may possibly require a custom mapping to your application's authorization scheme. For example, someone who is a "Manager" in Active Directory at Adatum might map to being an "Administrator" in Adatum's Tailspin Surveys application.

Chapter 6, "Securing Multi-Tenant Applications," of this guide discusses topics such as authentication and authorization in multi-tenant applications in more detail.

> *For more information about identity, authentication, and authorization in cloud applications see "A Guide to Claims-Based Identity and Access Control." You can download a PDF copy of this guide.*

The Command Query Responsibility Segregation (CQRS) Pattern

The CQRS pattern is an architectural design pattern that enables you to meet a wide range of architectural challenges such as managing complexity, managing changing business rules, or achieving scalability in some portions of your system. It's important to note that the CQRS pattern is not a top-level pattern, and should only be applied in the specific areas of a system where it brings clearly identifiable benefits.

Many of the multi-tenant considerations listed in this chapter relate to architectural challenges that CQRS can help you to address. However, you should not assume that multi-tenancy necessarily implies that you should use the CQRS pattern. For example, although the Tailspin Surveys application must be highly scalable to support many subscribers with different usage patterns, it is not an especially complex application. In particular, it is not a collaborative application where multiple users simultaneously edit the same data, which is one of the scenarios specifically addressed by the CQRS pattern. Furthermore, Tailspin does not expect the business rules in the Surveys application to change much over time.

> For more information about the CQRS pattern, and when you should consider using it, see the guide "A CQRS Journey."

Application Life Cycle Management Considerations

Your choice of a single-tenant or multi-tenant architecture will determine how easy it is to develop, deploy, maintain, and monitor your application.

Maintaining the Code Base

Many ISVs find that, when they move to hosting applications in the cloud instead of hosting them at client sites, release cycles become shorter. This means that they can much more quickly incorporate as part of a standard release a customization or enhancement requested by one client. This can benefit all subscribers while preventing unnecessary forking or multiple versions of the source code.

Maintaining separate code bases for different subscribers will rapidly lead to escalating support and maintenance costs for an ISV because it becomes more difficult to track which subscribers are using which version. This will lead to costly mistakes being made. A multi-tenant system with a single, logical instance guarantees a single code base for the application. If your multi-tenant application uses some single-tenant elements, there could be a short-term temptation (with long-term consequences) to branch the code in those elements for individual subscribers in order to meet the specific requirements of some subscribers.

In some scenarios, where there is a requirement for a high-degree of customization, multiple code bases may be a viable option but you should explore how far you can get with custom configurations or custom business rule components before going down this route. If you do need multiple code bases, you should structure your application such that custom code is limited to as few components as possible.

Handling Application Updates

A multi-tenant application that has a single code base makes it easy to roll out application updates to all your subscribers at the same time. This approach means that you have only a single logical instance to update, which reduces the maintenance effort. In addition, you know that all your subscribers are using the latest version of the software, which makes the support job easier. Windows Azure update domains facilitate this process by enabling you to roll out your update across multiple role instances without stopping the application. However, you must still carefully plan how you will perform a no-downtime update to your application, taking into consideration the fact that during the update process you may temporarily have instances with different versions of your software running simultaneously.

You should carefully test your update procedures before performing the update in your production environment, and ensure that you have a plan to revert to the original version if things don't go as planned.

If a client has operational procedures or software tied to a specific version of your application, any updates must be coordinated with that client. To mitigate the risks associated with updating the application, you can implement a rolling update program that updates some users, monitors the new version, and when you are confident in the new version, rolls out the changes to the remainder of the user base.

> *For information about how you can update a Windows Azure service and the different approaches that are available, see "Overview of Updating a Windows Azure Service."*

Monitoring the Application

Monitoring a single application instance is easier than monitoring multiple instances. In the multi-instance, single-tenant model any automated provisioning would need to include setting up the monitoring environment for the new instance, which will add to the complexity of the provisioning process for your application. Monitoring will also be more complex if you decide to use rolling updates because you must monitor two versions of the application simultaneously and use the monitoring data to evaluate the new version of the application.

Chapter 7, "Managing and Monitoring Multi-Tenant Applications," of this guide contains more information about implementing efficient management and monitoring practices for multi-tenant applications.

Using Third-Party Components

If you decide on a multi-tenant architecture, you must carefully evaluate how well any third-party components will work. You may need to take some additional steps ensure that a third-party component is "multi-tenant aware." With a single-tenant, multi-instance deployment, where you want to be able to scale out for large tenants, you will also need to verify that third-party components are "multi-instance aware."

Provisioning for Trials and New Subscribers

Provisioning a new client or initializing a free trial of your service will be easier and quicker to manage if it involves only a configuration change. A multi-instance, single-tenant model will require you to deploy a new instance of the application for every subscriber, including those using a free trial. Although you can automate this process, it will be considerably more complicated than changing or creating configuration data in a single-instance, multi-tenant application.

Chapter 7, "Managing and Monitoring Multi-Tenant Applications," of this guide contains more information about provisioning for new subscribers in multi-tenant applications.

Customizing the Application

Whether you choose a single-tenant or multi-tenant architecture, subscribers will still need to be able to customize the application.

Customizing the Application by Tenant

Subscribers will want to be able to style and brand the site for their own users. You must establish how much control subscribers will want in order to determine how best to enable the customization. This may vary from just the ability to customize the appearance of the application, such as by allowing subscribers to upload cascading style sheets and image files, to enabling subscribers to design complete pages that interact with the application's services through a standard API.

You can implement simple customizations through configuration values that tenants can change and that the application stores in its data store, such as custom logo images, welcome text, or switches to enable certain functionality. For example, in the Surveys application, subscribers can choose whether to integrate the application's identity infrastructure with their own infrastructure, and they can choose the geographic location for their surveys. This type of configuration data can easily be stored in Windows Azure storage.

Other applications may require the ability to enable users to customize the business process within the application to some degree. Options here would include implementing a plug-in architecture so that subscribers could upload their own code, or using some form of rules engine that enables process customization through configuration. To implement a plug-in architecture, you could consider hosting the PowerShell runtime (see *System.Management.Automation Namespace* on MSDN) in your application, or using the *Managed Extensibility Framework (MEF)*.

Customization is, of course, nothing new. Microsoft Dynamics CRM is a great example of an application that has these levels of customization available.

Allowing tenants to upload their own code increases the risks of introducing a security vulnerability or of application failure because you have less control over the code that is running in the application. Many Software as a Service (SaaS) systems apply limits to this. Most simply disallow it. Allowing tenants to upload code or scripts also increases the security risks associated with the application.

Another alternative to consider is enabling your application to call a service endpoint provided by the tenant, which performs some custom logic and returns a result.

You may also want to provide tenants with ways to extend the application without using custom code. Subscribers to the survey application may want to capture additional information about a survey respondent that the standard application does not collect. To achieve this you must implement a mechanism for customizing the UI to collect the data, and a way of extending the data storage schema to include the new data.

Chapter 7, "Managing and Monitoring Multi-Tenant Applications," of this guide contains more information about customizing applications for multiple tenants, and establishing an efficient on-boarding mechanism.

URLs to Access the Application

There are several different URL schemes that you could adopt in a multi-tenant application to enable tenants to access their data. The following describes some possible options for the Tailspin Surveys scenario where a subscriber can publish public surveys, and where public users do not need to sign in to access the survey:

If you need to provide extensive customizations per tenant, it may become impractical to handle this in a shared multi-tenant instance. In this case you should consider using a multi-instance, single-tenant architecture, but beware of the potential running costs of this approach and the difficulties in maintaining the application if there are multiple versions of the source code. This approach is usually appropriate only if you have a small number of tenants.

- **http://surveys.tailspin.com/{unique-survey-name}**. All surveys are available on the same domain, and subscribers must choose a unique survey name for every survey.

- **http://surveys.tailspin.com/{subscriber-name}/{survey-name}**. Again all surveys are made available on the same domain, but subscribers now only need to ensure that their own survey names are unique.

- **http://{subscriber-domain-name}.tailspinsurveys.com/{survey-name}**. Each subscriber is allocated its own unique sub domain, and subscribers only need to ensure that their own survey names are unique.

- **http://{subscriber-domain-name}/{survey-name}**. Each subscriber has its own domain, and subscribers only need to ensure that their own survey names are unique.

   Subscribers may prefer one of options where their company name is included in the URL that the public will access.

The following describes some possible options for the Tailspin Surveys scenario where a subscriber can design and manage their surveys, and where the subscriber must sign in:

- **https://surveyadmin.tailspin.com/**. After the subscriber logs on, they can design and manage their own surveys.
- **https://surveyadmin.tailspin.com/{subscriber-name}**. A subscriber must still log on to design and manage their surveys.
- **http://{subscriber-domain-name}.tailspinsurveys.com/admin**. Each subscriber is allocated its own unique sub domain, and the admin path requires the subscriber to log on.
- **https://{subscriber-domain-name}/**. Each subscriber has its own domain for accessing the administrative functionality, and subscribers must still log on.

In this case, the first option may be acceptable. There is no significant benefit in including the subscriber's name anywhere in the URL.

In both scenarios, as well as considering the preferences of the subscribers you also need to consider how the choice of URL scheme might affect other parts of the application, such as the provisioning process and the authentication mechanisms in use. For example, custom domain names take time to propagate through the Domain Name System (DNS). In addition, if you support multiple authentication schemes, you must consider how to select the correct authentication mechanism for a subscriber if the subscriber name is not included in the URL. If you need to use SSL, you also need to consider how to install the necessary certificates.

Chapter 4, "Partitioning Multi-Tenant Applications," discusses the URL scheme adopted by Tailspin to work with the two web roles in the Surveys application.

Financial Considerations

Your billing and cost model may affect your choice of single-tenant or multi-tenant architecture.

Billing Subscribers

For an application deployed to Windows Azure, Microsoft will bill you each month for the services (compute, storage, transactions, and so on) that each of your Windows Azure accounts consumes. If you are selling a service to your subscribers, such as in the Tailspin Surveys application, you need to bill your subscribers for the service.

Stress testing your application can help you to determine what resources you need to support a given number of tenants. This can help you decide how much to charge your subscribers.

Pay per Use Plans

One approach to billing is to use a pay-per-use plan. With this approach, you monitor the resources used by each of your subscribers, calculate the cost of those resources, and apply a markup to ensure you make a profit. If you use a single-tenant architecture and create a separate Windows Azure account for each of your subscribers, it's easy to determine how much an individual subscriber is costing in terms of compute time, storage, and so on, and then bill the subscriber appropriately.

However, for a single-tenant instance running in a separate Windows Azure account, some costs will effectively be fixed; for example, paying for a 24x7 compute instance or a Windows Azure SQL Database instance may make the starting cost too high for small subscribers. With a multi-tenant architecture, you can share the fixed costs between tenants, but calculating the costs per tenant is not so straightforward and you will have to add some additional code to your application to meter each tenant's application usage. Furthermore, subscribers will want some way of tracking their costs, so you will need to be transparent about how the costs are calculated and provide access to the captured usage data.

Fixed Monthly Fee Plans

A second approach is to adopt a billing model that offers the Surveys service for a fixed monthly fee. It is difficult to predict exactly what usage an individual subscriber will make of the service; for the Surveys application, Tailspin cannot predict how many surveys a subscriber will create or how many survey answers the subscriber will receive in a specified period. Therefore, the profit margin will vary between subscribers (and could even be negative in some cases).

By making Surveys a multi-tenant application, Tailspin can smooth out the differences in usage patterns between subscribers, making it much easier to predict total costs and revenue, and reduce the risk of taking a loss. The more subscribers you have, the easier it becomes to predict average usage patterns for a service.

From the subscriber's perspective, charging a fixed fee for the service means that subscribers know, in advance, exactly what their costs will be for the next billing period. This also means that you have a much simpler billing system. Some costs, such as those associated with storage and transactions, will be variable and will depend on the number of subscribers you have and how they use the service. Other costs, such as compute costs or the cost of a Windows Azure SQL Database instance, will effectively be fixed. To be profitable, you need to sell sufficient subscriptions to cover both the fixed and variable costs.

If your application can scale out and scale back automatically, it will directly affect your costs. Autoscaling can reduce your costs because it can help to ensure that you use just the resources you need. However, with an autoscaling solution you will also want to put some upper limits on the resources that your application can use to place a cap on potential costs.

Different Levels of Fixed Fees

If your subscriber base is a mixture of heavy users and light users, a standard monthly charge may be too high to attract smaller users. In this scenario, you will need a variation on the second approach to offer a range of packages for different usage levels. For example, in the Surveys application, Tailspin might offer a light package at a lower monthly cost than the standard package. The light package may limit the number of surveys a subscriber can create or the number of survey responses that a subscriber can collect each month.

Offering a product where different subscribers can choose different features and/or quotas requires that you design the product with that in mind. Such a requirement affects the product at all levels: presentation, logic, and data. You will also need to undertake some market research to determine the expected demand for the different packages at different prices to try to estimate your expected revenue stream and costs.

Managing Application Costs

You can divide the running costs of a Windows Azure application into fixed and variable costs. For example, if the cost of a compute node is $0.12 per hour, the cost of running two compute nodes (to gain redundancy) 24x7 for one month is a fixed cost of approximately $180. If this is a multi-tenant application, all the tenants share that cost. To reduce the cost per tenant you should try to have as many tenants as possible sharing the application, without causing a negative impact on the application's performance. You also need to analyze the application's performance characteristics to determine whether scaling up by using larger compute nodes or scaling out by adding additional instances would be the best approach when demand increases. Chapter 5, "Maximizing Availability, Scalability, and Elasticity," discusses the pros and cons of scaling out by adding more instances compared to scaling up by using larger instances.

Variable costs will depend on how many subscribers you have, and how those subscribers use the application. In the Tailspin Surveys application, the number of surveys and the number of respondents for each survey will largely determine monthly storage and transaction costs. Whether your application is single-tenant or multi-tenant will not affect the cost per tenant; regardless of the model, a specific tenant will require the same amount of storage and use the same number of compute cycles. To manage these costs, you must make sure that your application uses these resources as efficiently as possible.

> *For more information about estimating Windows Azure costs, see Chapter 6, "Evaluating Cloud Hosting Costs" in the guide "Moving Applications to the Cloud." You can find information about storage costs by using the Windows Azure Pricing calculator, and in the blog post "Understanding Windows Azure Storage Billing – Bandwidth, Transactions, and Capacity."*

Engineering Costs

Windows Azure bills you for the cost of running your application in the cloud. You must also consider the costs associated with designing, implementing, and managing your application. Typically, multi-tenant application elements are more complex than single-tenant application elements. For example, you must consider how to isolate the tenants within an instance so that their data is kept private, and consider how the actions of one tenant might affect the way the application behaves for other tenants. This additional complexity can add considerably to the engineering costs associated with both building a multi-tenant application, and managing it.

MORE INFORMATION

All links in this book are accessible from the book's online bibliography available at: *http://msdn.microsoft.com/library/jj871057.aspx.*

For more information about working with the Windows Azure platform including planning, designing, and managing, see *"Windows Azure Developer Guidance."*

For more information about designing multi-tenant applications for Windows Azure, see *"Designing Multitenant Applications on Windows Azure."*

For more information about the costs associated with running a Windows Azure application, see *"Estimating Cost Of Running The Web Application On Windows Azure"* and *"Windows Azure Cost Assessment."*

For more information about ALM considerations, see *"Testing, Managing, Monitoring and Optimizing Windows Azure Applications."*

For more information about continuous delivery and using Team Foundation Service with Windows Azure, see *"Continuous Delivery for Cloud Services in Windows Azure."*

3

Choosing a Multi-Tenant Data Architecture

This chapter discusses important factors you must consider when designing the data architecture for multi-tenant applications, and explores how the Tailspin Surveys application uses data. It describes the data model used by the Surveys application, and then discusses why the team at Tailspin chose this data model with reference to a number of specific scenarios in the application. Finally, it describes how and why the application also uses Windows Azure SQL Database.

STORING DATA IN WINDOWS AZURE APPLICATIONS

Windows Azure offers several options for storing application data. In a multi-tenant application your data architecture typically requires you use a partitioning scheme that ensures each tenant's data is isolated, and that the application is scalable. In addition, you may need to consider how to make your storage solution extensible in order to support per tenant customization.

There are many factors to consider when selecting the type of storage to use in your application, such as features, cost, supported programming models, performance, scalability, and reliability. This section outlines the main options that are available, and identifies the key features that relate specifically to multi-tenancy. For a more general discussion of the pros and cons of the data storage options, see the associated patterns & practices guide *"Moving Applications to the Cloud."*

Windows Azure table storage is often referred to as schema-less because every entity in a table could have different set of properties. However, when all the entities in a table have the same set of properties (they have the same schema) a Windows Azure table is much like a table in a traditional database.

Windows Azure Table Storage

Windows Azure tables contain large collections of state stored as property bags. Each property bag is called an entity, and each entity in a table can contain a different set of properties. You can filter and sort the entities in a table.

Each table can be subdivided into partitions by using a partition key, and the scalability of a solution that uses Windows Azure table storage is primarily determined by the use of appropriate partition keys. Searching for and accessing entities that are stored in the same partition is much faster than searching for and accessing entities across multiple partitions, and a query that specifies a partition key and row key is typically the most performant. In addition, Windows Azure table storage only supports transactions across entities that reside on the same partition.

The choice of keys can also help to define your multi-tenant data architecture. In a multi-tenant application you typically need to access only the data associated with a single tenant in a query, so you should use partition keys based on tenant IDs. You can store different entity types, such as tenant header and detail records, in the same table partition. Therefore, any queries that combine data from different entities associated with the same tenant will run efficiently.

Windows Azure tables are associated with a Windows Azure storage account that is accessed using an account key, and each storage account is tied to a specific Windows Azure data center.

Partitioning by tenant is a very natural boundary to choose in a multi-tenant application. Almost all of your queries will be scoped to a single tenant.

Windows Azure Blob Storage

Windows Azure blob storage is for storing individual items such as documents, media items, XML data, or binary data. Blobs are ideal for unstructured data, so each tenant can easily store any customized data in blob storage.

You place blobs in containers, which you can use to control the visibility of your blobs. Windows Azure blobs and blob containers are associated with a Windows Azure storage account that is accessed using an account key.

> *A single storage account can contain tables, blobs, and queues. You can have multiple storage accounts in a Windows Azure subscription.*

Windows Azure SQL Database

Windows Azure SQL Database is a scalable, relational database management system for Windows Azure. It is based on SQL Server and is very similar in functionality. It is made available using the Platform as a Service (PaaS) model, so you are billed based on your usage.

SQL Database also supports federation to enable greater scalability. Federation makes use of a technique called sharding that splits tables horizontally by row across multiple databases. This allows you to take advantage of the database resources in the cloud on demand, remove the risk of having a single point of failure, and minimize I/O bottlenecks and database throttling. For a detailed discussion of SQL Database Federation see the article *"Scaling Out with SQL Azure Federation"* in MSDN Magazine.

Other Storage Options

Other storage options for a Windows Azure application include running a relational database such as SQL Server or MySQL in a Windows Azure Virtual Machine (VM), or running a no-SQL database such as MongoDB in a Windows Azure VM.

Storage Availability

One additional aspect you should consider when choosing a storage method is availability. Storage availability is mainly governed by two factors: whether the storage mechanism responds without fail to every request, and whether the behavior of the network connection between the application and storage constrains or even prevents access. Application performance and user experience will suffer if there is a problem when accessing storage, even if it is only a short delay while the attempt is retried, although this can be minimized in some circumstances by the judicious use of caching.

Each type of data store has a guaranteed availability and a specific maximum throughput. For example, the Service Level Agreement (SLA) for Windows Azure storage indicates that the expected availability is 99.9% (anything less than this will result in a partial refund of hosting cost). Windows Azure blob storage has a throughput limit of 60 MB per second for a single blob, while Windows Azure table storage has a throughput limit of 500 entities per second for a single partition. Careful storage design and the use of appropriate partitioning strategies can help to avoid hitting these limits.

Windows Azure SQL Database availability is also guaranteed to be 99.9%, though actual response time may be affected by throttling that is applied automatically when the underlying system detects overloading of the database server or a specific database. Throttling is applied initially by increasing response times for the affected database. However, if the situation continues Windows Azure will begin to refuse connections until the load decreases. Good query design, promptly closing connections, and the appropriate use of caching can minimize the chances of encountering database throttling.

Microsoft SQL Server 2012 supports new availability features through AlwaysOn Availability Groups. You can configure multiple SQL Server 2012 instances in Windows Azure Virtual Machines as an availability group to provide instant failover, plus the capability to read from replicas as well as from the primary instance. You can also use both a synchronous and an asynchronous commit approach to maximize performance and availability. The SLA for hosted service roles guarantees an uptime of 99.9% as long as two or more roles are deployed, and 99.9% availability of connectivity to the roles.

The second main factor that affects storage availability is the performance of the network between the application and the data source. Application and data store should, wherever possible, be located in the same datacenter to minimize network latency. The use of affinity groups can also help by causing the resources to be located in the same sector of the datacenter.

Where the application and the data it uses must be geographically separated, consider using replicas of the primary data store at the same location as the application. The geo-replication feature of Windows Azure storage can create multiple copies of the data in several datacenters, but you cannot specify which datacenters are used. However, you can create storage accounts in the appropriate datacenters and copy the data between them, and use Windows Azure Caching or the Content Delivery Network (CDN) to cache data at locations closer to the application.

When using SQL Database or SQL Server consider placing database replicas in locations close to the application, and using SQL Database Sync to synchronize the data.

> You can download the SLAs for Window Azure services from *"Service Level Agreements."* For more information about maximizing performance and availability in multi-tenant applications see Chapter 5, *"Maximizing Availability, Scalability, and Elasticity."*

MULTI-TENANT DATA ARCHITECTURES

Your data architecture must ensure that a subscriber's data is kept private from other subscribers, and that your solution is scalable. Your application may also need to support customized data storage.

> For more information about multi-tenant data architectures, see *"Multi-Tenant Data Architecture"* and *"Architecture Strategies for Catching the Long Tail."*

Partitioning to Isolate Tenant Data

The perceived risk of either accidental or malicious data disclosure is greater in a multi-tenant model. It will be harder to convince subscribers that their private data is safe if they know they are physically sharing the application with other subscribers. However, a robust design that logically isolates each tenant's data can provide a suitable level of protection. This type of design might use database schemas where each tenant's tables are in a separate schema, database security features that enable you to use access control mechanisms within the database, a partitioning scheme to separate tenants' data, or a combination of these approaches.

> For a more detailed exploration of data security issues in multi-tenant applications, see Chapter 6, *"Securing Multi-Tenant Applications,"* of this guide.

In all multi-tenant applications the design must ensure that tenants can access only their own data. To achieve proper isolation you must be sure not to reveal any storage account keys, and be sure that all queries in your code access and return the correct tenant's data.

For all the data storage mechanisms described in this section, if the tenant provides the subscription this makes clear that the tenant owns, and is responsible for, the data stored in any storage account or database in the subscription.

The following table shows the partitioning schemes you could use based on Windows Azure subscriptions. These partitioning schemes can be used with Windows Azure storage, SQL Database, and hosting a database in a VM.

Partitioning scheme	Applies to	Notes
One subscription per tenant	Table storage Blob storage SQL Database Database hosted in VM	Makes it easy to bill individual tenants for the storage resources they consume. Enables tenants to provide their own storage and then manage it. During the provisioning process, the tenant would need to provide access details such as storage account keys or database passwords to the provider. You need to be careful about the location of the storage account in relation to the location of the cloud services roles to control data transfer costs and minimize latency. Provisioning a new Windows Azure subscription is a manual process.
Group multiple tenants in a subscription	Table storage Blob storage SQL Database Database hosted in VM	If your tenants can subscribe to different levels of functionality for the application (such as light, standard, and premium), using a different subscription for each level but grouping all the tenants for each level in the same subscription makes it easier to track the costs of providing each level of functionality. You must still partition the data that belongs to different tenants within a subscription using one of the other partitioning schemes.

The following table shows the partitioning schemes you could use with Windows Azure storage, in addition to partitioning by Windows Azure subscription.

Partitioning scheme	Applies to	Notes
One storage account per tenant	Table storage Blob storage	Five storage accounts per subscription is a soft limit. You can request to extend this up to the hard limit of 20 storage accounts per subscription; however, this may limit the usefulness of this partitioning approach. Each storage account appears as a line item on your Windows Azure bill, so this approach can be useful if you want to identify the precise costs per tenant.
Group multiple tenants in a storage account	Table storage Blob storage	Enables you to group tenants by geographic region, by regulatory requirements, and by replication requirements. You must still partition the data that belongs to different tenants within a storage account using one of the other partitioning schemes.
One table per tenant	Table storage	There is no practical limit to the number of tables you can have in a Windows Azure storage account. You can automate creating a table within the provisioning process. Include the tenant's ID in the table name.
Single table with one partition key per tenant	Table storage	There is no practical limit on the number of partitions in a table. Include the tenant's ID in the partition key.
One container per tenant	Blob storage	There is no practical limit on the number of containers you can have in a Windows Azure storage account. Enables you to store all the blobs associated with a single tenant in a single container, much like using a folder on the file system. This makes it easy to manage tenant specific data. For example, provisioning and de-provisioning tenants, backup, archiving, and setting access policies. You can create containers automatically during the provisioning process. Include the tenant's ID in the container name.
Blob naming convention	Blob storage	There is no practical limit to the number of blobs you can have in a container. Include the tenant's ID in the blob name whenever you create a new blob.

Windows Azure storage is billed by the amount of storage used and by the number of storage transactions, so from a cost perspective it doesn't matter how many separate storage accounts or containers you have.

Using blob containers instead of a blob naming convention is the simplest solution to identify who the blob belongs to.

The following table shows the partitioning schemes you could use with Windows Azure SQL Database, or with a database such as SQL Server or MySQL hosted in a Windows Azure VM. These are in addition to partitioning by Windows Azure subscription (as described in the previous tables).

Partitioning scheme	Applies to	Notes
One server per tenant	SQL Database	Each SQL Database server can be hosted in a different geographic region.
		There is a limit on the number of SQL Database servers for a subscription.
One VM per tenant	Database hosted in VM	For each tenant you will incur the costs associated with running a Windows Azure VM.
		You can give the tenant access to the VM.
		You are limited to 25 VMs in a Windows Azure IaaS deployment.
One database per tenant	SQL Database / Database hosted in VM	Each database you create has a specified size that determines the monthly cost.
		You can create each database in a different logical server, or host several databases in a single server.
		You have more databases to manage.
		There is a limit on the number of databases you can install on a SQL Database server.
Multiple tenants per database with per tenant tables	SQL Database / Database hosted in VM	Enabling multiple tenants to share a database helps to reduce per tenant costs by efficiently using the storage you are paying for.
		You isolate tenant data by using separate tables for each tenant. You can use a naming convention that includes the tenant ID in the table name, or use a different database schema for each tenant.
Multiple tenants per database with shared tables	SQL Database / Database hosted in VM	You must have a partitioning scheme to identify each tenant's records in each table, such as using the tenant's ID as part of the key.

For more information about managing multi-tenant data in relational databases such as SQL Server or SQL Database, see the article *"Multi-Tenant Data Architecture"* and the blog post *"Full Scale 180 Tackles Database Scaling with Windows Azure."*

Windows Azure SQL Database cost is based on the number and size of databases you have, so it makes sense from a cost perspective to have as many tenants as possible sharing each instance.

At the time of writing you are limited to six Windows Azure SQL Database servers per Windows Azure subscription and 150 databases per server, although these limits may be extended on request and may change in the future.

Shared Access Signatures

Both Windows Azure table storage and Windows Azure blob storage support shared access signatures as a mechanism for controlling access to data. You can use shared access signatures to ensure isolation of tenant data.

Typically, all table data in a storage account is available for read and write access to any client that has access to the storage account key. With blob storage, a client that has access to the storage account key also has read and write access to all blobs in all containers in the storage account. In addition, you can grant public read access to a blob so that anyone who knows the blob's URL will be able to read its content.

A shared access signature is a way to grant temporary access to a resource using a token. Your application can generate a shared access signature token for a blob container, an individual blob, or for a range of entities in a table. A shared access signature grants the holder of the token specific access rights such as read, write, update, and delete for a fixed time. You could use a role instance to generate shared access signatures for a specific tenant's data and then issue those tokens to another role instance, possibly in another Windows Azure subscription. In this way only specific roles need to have access to the storage account keys that grant full access to the data in the storage account.

For more information about shared access signatures see Chapter 6, "Securing Multi-tenant Applications," in this guide, and the blog post *"Introducing Table SAS (Shared Access Signature), Queue SAS and update to Blob SAS."*

> SQL Federation, described later in this chapter as a technique for scaling your SQL Database instance both in size and performance, uses the "multiple tenants per database with shared tables" approach to enable it to scale out your database. To support a multi-tenant application, SQL Federation typically uses the tenant ID to determine which database instance in the federation should store a particular tenant's record. SQL Federation also supports filtered connections; you can use these to isolate tenant data and ensure that only individual tenant's data can be accessed over a connection.

> You are billed for Windows Azure SQL Database based on the number of databases you have, and the size of the databases. If you transfer data in and out of Windows Azure SQL Database from within the same data center there's no data transfer cost, but if you transfer out of the data center you'll be charged for the data transfer.

Data Architecture Extensibility

There are a number of ways you can design your data storage to enable tenants to extend the data model so that it includes their own custom data. These approaches range from each tenant having a separate schema, to providing a set of pre-defined custom columns, to more flexible schemas that enable a tenant to add an arbitrary number of custom fields to a table.

If you use Windows Azure SQL Database, much of the application's complexity will result from having to work within the constraints of fixed data schemas. If you are using Windows Azure table storage, the complexity will arise from working with variable schemas. Windows Azure table storage allows records in the same table to have completely different structures, which allows for a great deal of flexibility at the cost of more complexity in the code.

> *Microsoft SharePoint is an example of an application with a fixed schema database that looks extremely flexible.*

Custom extensions to the application's data model should not require changes to the application code. To enable the application's business logic and presentation logic to integrate with the data model extensions, you will require either a set of configuration files that describe the extensions, or write code that can dynamically discover the extensions. However, if you enable tenants to extend the application through some predefined extension points or through an API, an extension could include both changes to the data model and to the code.

The following table summarizes the options for implementing an extensible data architecture in Windows Azure table storage:

You should aim to have a single codebase for your application, and avoid the situation where custom data extensions require different codebases.

Extensibility approach	Notes
Separate table per tenant	Each table can use custom schemas for that particular tenant.
Single table with multiple schemas	Each tenant can use custom schemas for the entities it stores in the table.
Single schema with separate tables holding custom data	In Windows Azure table storage, transactions are only supported within a partition on a table. With this approach it is not possible to save all the data associated with an entity in a single transaction.

Figure 1 illustrates these alternatives using two of Tailspin's subscribers, Adatum and Fabrikam, as examples. Each subscriber is storing different data as part of a survey definition.

Separate table per tenant

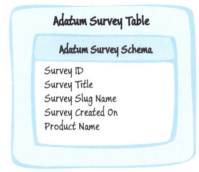

Adatum Survey Table

Adatum Survey Schema

Survey ID
Survey Title
Survey Slug Name
Survey Created On
Product Name

Fabrikam Survey Table

Fabrikam Survey Schema

Survey ID
Survey Title
Survey Slug Name
Survey Created On
Campaign ID
Owner

Single table with multiple schemas

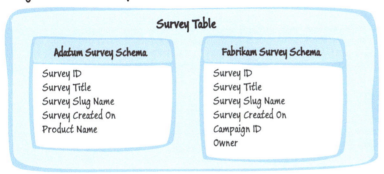

Survey Table

Adatum Survey Schema

Survey ID
Survey Title
Survey Slug Name
Survey Created On
Product Name

Fabrikam Survey Schema

Survey ID
Survey Title
Survey Slug Name
Survey Created On
Campaign ID
Owner

Single schema with separate table holding custom data

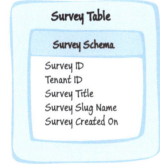

Survey Table

Survey Schema

Survey ID
Tenant ID
Survey Title
Survey Slug Name
Survey Created On

Adatum Extension Table

Adatum Extension Schema

Survey ID
Product Name

Fabrikam Extension Table

Fabrikam Extension Schema

Survey ID
Campaign ID
Owner

FIGURE 1
Examples showing per tenant customizations

A slug name stored in the Survey table is a string where all whitespace and invalid characters are replaced with a hyphen (-). The term comes from the newsprint industry and has nothing to do with those things in your garden!

Using different schemas for different tenants—either in the same table or in different tables—enables a great deal of flexibility for extending and customizing the data architecture. In Windows Azure table storage you don't need to predefine the schemas you will use before adding entities to a table. However, managing multiple schemas will add to the complexity of your solution. By limiting the customizability of your application you can limit the complexity of your solution.

The following table summarizes the options for implementing an extensible data architecture in SQL Database, and in relational databases such as SQL Server and MySQL that can run in a Windows Azure VM.

Extensibility approach	Notes
Separate database with custom schema per tenant	Each database can use a different schema to accommodate the requirements of each tenant. Typically, the custom schema must be defined during the provisioning process.
Shared database with separate schema or tables for each tenant	For relational databases that support multiple schemas within a database, each tenant can use a custom schema. Otherwise each tenant can have its own set of tables, identified using a naming convention. Typically, the custom schema must be defined during the provisioning process.
Single fixed schema with a set of columns available for custom data	Limits the amount of customization that is possible because there are a limited number of custom columns available.
Single fixed schema with separate tables holding custom data	Allows slightly more flexibility than using custom columns.

Using custom schemas will add to the complexity of the solution, especially because the schema must be defined before you can use the database. It is difficult to change a schema after you have added data to a table.

Data Architecture Scalability

If you can partition your data horizontally you will be able to scale out your data storage. In the case of Windows Azure SQL Database, if you decide that you need to scale out you should be able to move all of an individual tenant's data to a new database instance. The partitioning scheme you choose will also affect the scalability of your solution.

For Windows Azure table storage, the most significant decision that affects scalability is the choice of partition key for a table. Queries that operate on a single partition are much more efficient than queries that access entities that exist on multiple partitions. In addition, you can only use transactions when all the entities involved reside on the same partition in the same table. Typically, a partition key that includes the tenant ID will help to make your Windows Azure table storage design scalable because the majority of your queries will need to access only a single partition to retrieve their data. For more information, see Chapter 7, *"Moving to Windows Azure Table Storage"* of the related patterns & practices guide *"Moving Applications to the Cloud."*

For SQL Database, federation helps you to scale out across multiple databases by partitioning your data horizontally. If you decide to partition your tenant data by including the tenant ID in the primary key, this can be combined with SQL Database federation to achieve scalability.

> *Partitioning data horizontally, also known as sharding, implies taking some of the records in a table and moving them to a new table. Partitioning data vertically implies taking some fields from every row and placing them in a different table. For a discussion of federation and sharding in Windows Azure SQL Database, see "Federations in Windows Azure SQL Database."*

An Example

This section shows a set of alternative data architectures in order to illustrate some of the key issues you should consider, such as isolation, extensibility, and scalability. This simple example makes the following assumptions about the application and the data:

- A multi-tenant application stores the data.
- You are storing the data in Windows Azure table storage.
- There are two basic record types: a header record and a detail record where there is a one-to-many relationship between them.
- All queries in the application access records for a specific month in a specific year.
- Tenants with a premium subscription use an extended version of the detail record. Tenant B is an example of a tenant with a premium subscription; tenant A has a standard subscription.

The following is a list of five alternatives, and it describes the entity types stored in each table in each case. This is not an exhaustive list of the possible options, but it does illustrate a range of possibilities that you might consider. All of the options are designed to ensure that the application can keep each tenant's data isolated. You can find a discussion of some of the advantages and limitations of the different approaches at the end of this section.

Option 1 — Using a Single Table

Application Data

Partition Key	Row Key	Entry Type
Tenant ID, Month, Year	Header Entity ID	Header record
Tenant ID, Month, Year	Detail Entity ID	Detail record (standard schema)
Tenant ID, Month, Year	Detail Entity ID	Detail record (extended schema)

Option 2 — Table per Tenant

Tenant A (uses standard detail record schema)

Partition Key	Row Key	Entry Type
Month, Year	Header Entity ID	Header record
Month, Year	Header Entity ID, Detail Entity ID	Detail record (standard schema)

Tenant B (uses extended detail record schema)

Partition Key	Row Key	Entry Type
Month, Year	Entity ID	Header record
Month, Year	Header Entity ID, Detail Entity ID	Detail record (extended schema)

Option 3 — Table per Base Entity Type

Header Records

Partition Key	Row Key	Entry Type
Tenant ID, Month, Year	Header Entity ID	Header record

Detail Records

Partition Key	Row Key	Entry Type
Tenant ID, Month, Year	Header Entity ID, Detail Entity ID	Detail record (standard schema, standard tenants)
Tenant ID, Month, Year	Header Entity ID, Detail Entity ID	Detail record (extended schema, premium tenants)

Option 4 — Table per Entity Type

Header Records

Partition Key	Row Key	Entry Type
Tenant ID, Month, Year	Header Entity ID	Header record

Detail Records (standard tenants)

Partition Key	Row Key	Entry Type
Tenant ID, Month, Year	Header Entity ID, Detail Entity ID	Detail record (standard schema, standard tenants)

Detail Records (premium tenants)

Partition Key	Row Key	Entry Type
Tenant ID, Month, Year	Header Entity ID, Detail Entity ID	Detail record (extended schema, premium tenants)

Option 5 — Table per Entity Type per Tenant

Tenant A Header Records

Partition Key	Row Key	Entry Type
Month, Year	Header Entity ID	Header record

Tenant B Header Records

Partition Key	Row Key	Entry Type
Month, Year	Header Entity ID	Header record

Tenant A Detail Records (standard schema)

Partition Key	Row Key	Entry Type
Month, Year	Header Entity ID, Detail Entity ID	Detail record (standard schema, standard tenants)

Tenant B Detail Records (extended schema)

Partition Key	Row Key	Entry Type
Month, Year	Header Entity ID, Detail Entity ID	Detail record (extended schema, premium tenants)

Comparing the Options

There is no right or wrong choice from the options listed above; the specific requirements of your application will determine which one you choose. There are many considerations that might affect your choice, some of which include the following:

- **Transactional behavior**. Windows Azure table storage only supports transactions within a partition. If there is a requirement to support transactions that span the header and detail records, options one and two provide this functionality.
- **Query Performance**. Queries against Windows Azure table storage perform best when you can specify the partitions that contain the data in the query. You need to analyze the queries in your application to decide on a partition scheme that can optimize their performance.
- **Code complexity**. Dealing with multi-schema tables is more complex than single-schema tables. However, if you plan to have additional schema extensions or allow per tenant schema customizations, it could be more complex to manage many different tables in addition to supporting multiple schemas.
- **Managing the data**. Performing management operations such as backing up, creating and deleting tenants, and logging may be easier if each tenant has its own set of tables.
- **Scale out**. For very large volumes of data and transactions, you may want to scale out to use multiple Windows Azure storage accounts. You should chose an architecture that makes it easy to divide your data across storage accounts, most likely by placing some tenants in one account and others in a different one. For more information about scaling multi-tenant applications, see Chapter 5, "Maximizing Availability, Scalability, and Elasticity."
- **Geo location**. To reduce latency and improve performance you may want to store the data belonging to a particular tenant in a particular datacenter. Again, your architecture should support this type of partitioning.

The options shown above illustrate alternative approaches to storing multi-tenant data and do not specifically address the issue of scalability. There is an anti-pattern for Windows Azure table storage where you only append or prepend entities to a specific partition: all writes then go to the same partition, limiting the scalability of the solution. A common way to implement this anti-pattern is to use the current date as the table partition key, so in the options shown above you should verify whether the anticipated volume of transactions means that a choice of month and year for the partition key is sub optimal. For more information, see the presentation *Windows Azure Storage Deep Dive* on Channel 9.

GOALS AND REQUIREMENTS

This section describes the specific goals and requirements that Tailspin has for the Surveys application with respect to the architecture and design of the data storage elements of the application.

Isolation of Tenants' Data

Tailspin wants to ensure that each tenant's data is fully isolated from every other tenant's data. When a tenant publishes a survey it becomes publically available, but each tenant must be able to manage its survey definitions privately. For example, a tenant can control when its surveys become publically visible. In addition, the survey response data must be confidential to the tenant who created the survey.

Application Scalability

As Tailspin signs up more tenants for the Surveys application it must be able to scale out the storage. This is particularly important for Survey response data because some Surveys might result in a large number of responses. Furthermore, Tailspin would like to be able to scale the Surveys application out (and in) automatically because Tailspin cannot predict when a tenant will create a survey that elicits a large number of responses.

Extensibility

As a feature for premium subscribers, Tailspin plans to allow tenants to store additional, tenant-specific information as part of each survey. Tenants must be able to store one or more custom fields as part of a survey definition. Tenants can use this information to store metadata for each survey, such as the product the survey relates to or the person who owns the survey.

Tailspin plans to add support for new question types in the future. However these will be made available to all subscribers and will not be treated as a per tenant customization.

Paging through Survey Results

The owner of a survey must be able to browse through the survey results, displaying a single survey response at a time. This feature is in addition to being able to view summary statistical data, and being able to analyze the results using Windows Azure SQL Database. The Surveys application contains a Browse Responses page for this function.

The design of this feature of the application must address two specific requirements. The first requirement is that the application must display the survey responses in the order they were originally submitted. The second requirement is to ensure that this feature does not adversely affect the performance of the web role.

Adding support for a new question type will affect many areas of Tailspin Surveys such as storage, the UI, exporting to SQL Azure, and data summarization. Therefore, Tailspin will develop any new question types as extensions available to all subscribers.

Exporting Survey Data to SQL Database for Analysis

The Surveys application uses Windows Azure storage to store survey definitions and survey responses. Tailspin chose to use Windows Azure storage because of its lower cost and because the cost depends on the amount of usage—both in terms of capacity used and the number of storage transactions per month. However, to control the costs associated with storage, the Surveys application does not offer a great deal of flexibility in the way that subscribers can analyze the survey responses. A subscriber can browse through the responses to a survey in the order that users submitted their responses, and a subscriber can view a set of fixed design summaries of the statistical data for each survey.

Windows Azure SQL Database allows subscribers to perform complex analysis on their survey results. Subscribers can also create custom reports using Windows Azure SQL Reporting. For complex datasets subscribers can use a Windows Azure Big Data solution based on the Apache Hadoop software library.

To extend the analysis capabilities of the Surveys application, Tailspin allows subscribers to dump their survey responses into a Windows Azure SQL Database instance. They can then perform whatever detailed statistical analysis they want using tools of their choosing, or they can use this as a mechanism to download their survey results to an on-premise application by exporting the data from Windows Azure SQL Database.

The application must be able to export all survey data to Windows Azure SQL Database, including the question definitions as well as the survey responses.

This feature is included in the monthly fee for a Premium subscription. Subscribers at other levels can purchase this feature as an add-on to their existing subscription. Subscribers who choose to use this feature have their own private instance of Windows Azure SQL Database to ensure that they are free to analyze and process the data in any way that they see fit. For example, they may choose to create new tables of summary data, design complex data analysis queries, or design custom reports. This data must be kept confidential for each tenant.

Overview of the Solution

This section describes some specific features of the Tailspin Surveys application related to data architecture, and discusses the reasons for the solution adopted by Tailspin. It identifies any alternatives Tailspin considered and the trade-offs they imply.

Storage Accounts

Tailspin considered using separate Windows Azure subscriptions for its premium and standard subscribers, but the additional complexity of managing separate subscriptions did not provide any significant benefits. The billing information provided by Windows Azure for a single subscription gives Tailspin enough detailed information to understand its costs.

Tailspin does plan to use separate storage accounts for the different regions where it hosts the Surveys application. This is discussed in more detail in Chapter 5, "Maximizing Availability, Scalability, and Elasticity."

> *A storage account is tied to a specific Windows Azure data center, so Tailspin must use multiple storage accounts to store data in different regions.*

The Surveys Data Model

This section describes the data model in the Surveys application and explains how the table design partitions the data by tenant.

The Surveys application uses a mix of table storage and blob storage to store its data. The section, "Options for Saving Survey Responses" in Chapter 5, "Maximizing Availability, Scalability, and Elasticity," and the following section in this chapter discuss why the application uses blob storage for some data. Figure 2 shows, at a high level, which data is stored in the different types of storage.

The Surveys application uses blob and table storage.

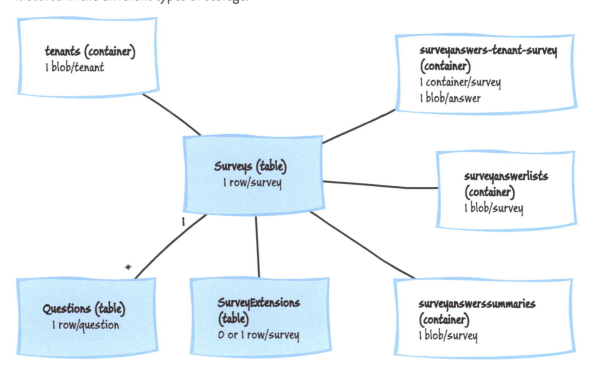

FIGURE 2
Data storage in the Surveys application

Storing Survey Definitions

The Surveys application stores the definition of surveys in two Windows Azure tables. This section describes these tables and explains why Tailspin adopted this design.

Tailspin chose to store survey definition in Windows Azure table storage to simplify the implementation of this part of the application. Every survey has some header data that describes the survey, and an ordered list of questions. It is easy to model this using a parent/child relationship between the two types of entity stored in the tables.

Survey definitions are read whenever a user responds to a survey, and Tailspin minimizes the number of storage transactions involved by caching survey definitions in memory. Although Tailspin cannot use a transaction when it saves survey definitions because it is using two tables, it maintains data consistency by saving the question entities before the corresponding survey entity. This can lead to orphaned question entities if a failure occurs before the survey entity is saved and so Tailspin will regularly scan the Questions table for such orphaned rows.

> A transaction cannot be used to ensure full consistency of data when saving it in two separate Windows Azure storage tables. By saving the data in the child table first, and then saving the linked row in the parent table only if the child rows were saved successfully, Tailspin can prevent inconsistencies in the data. However, it does mean that some child rows could be saved without a matching row in the parent table if the second save operation fails, and so Tailspin will need to create and run a separate process that periodically sweeps the child table for orphaned rows. This will add to the development and runtime cost.

The following table describes the fields in the Surveys table. This table holds a list of all of the surveys in the application.

Field name	Notes
PartitionKey	This field contains the tenant name. Tailspin chose this value because they want to be able to filter quickly by tenant name, and ensure the isolation of survey definitions by tenant.
RowKey	This field contains the tenant name from the **PartitionKey** field concatenated with the slug name version of the survey name. This makes sure that a subscriber cannot create two surveys with the same name. Different subscribers can use the same name for a survey; these surveys are differentiated by the tenant name part of the ID.
Timestamp	Windows Azure table storage automatically maintains the value in this field.
CreatedOn	This field indicates when the subscriber created the survey. This will differ from the **Timestamp** value if the subscriber edits the survey.
SlugName	The slug name version of the survey name.
Title	The survey name.

The following table describes the fields in the Questions table. The application uses this table to store the question definitions and to render a survey.

Field name	Notes
PartitionKey	This field contains the row key from the Surveys table, which is the tenant name from the **PartitionKey** field in the Surveys table concatenated with the slug name version of the survey name. This enables the application to insert all questions for a survey in a single transaction and to retrieve all the questions in a survey quickly from a single partition.
RowKey	This field contains a formatted tick count concatenated with the position of the question within the survey. This guarantees a unique **RowKey** value and defines the ordering of the questions.
Timestamp	Windows Azure table storage automatically maintains the value in this field.
Possible-Answers	This field contains a list of the possible answers if the question is a multiple-choice question.
Text	The question text.
Type	The question type: Simple text, multiple choice, or five stars (a numeric range).

Remember that Windows Azure table storage only supports transactions within a single partition in a single table, and that a transaction cannot modify more than 100 rows at a time.

Each of these tables uses the tenant ID in the partition key. This helps to isolate tenant data because all of the queries in the Tailspin Surveys application include a partition key value. Using a single table to store all of the tenants' data makes it easier for Tailspin to manage this data. For example, Tailspin can easily back up all of the survey definitions.

Premium tenants can add their own custom metadata to enable linking with their own applications and services. This level of customization requires Tailspin to extend the schema of the Surveys table in Windows Azure storage and to add code to the application that recognizes this custom data. To extend the survey table schema Tailspin considered two alternative approaches:

• Store the custom data in the existing survey table, using different custom fields for each tenant.

• Store the custom data in a separate table.

Tailspin chose the first option. Windows Azure table storage allows you to use multiple schemas in the same table; therefore, each tenant can have its own custom fields. Furthermore, if a tenant changes the type of custom data that it needs to store, it can itself have multiple schemas. The following table illustrates how Adatum added a new custom field before it added its second survey, and how Fabrikam has different custom fields from Adatum.

Partition key	Row key	Slug name	Title	Product name	Owner	Promotion
adatum	adatum_survey-1	survey-1	Survey 1	Widgets		
adatum	adatum_survey-2	survey-2	Survey 2	Gadgets	Mary	
fabrikam	fabrikam_survey-1	survey-1	Survey 1			Promo 1
fabrikam	fabrikam_survey-2	survey-2	Survey 2			Promo 2

The Surveys table is a multi-schema table. Each tenant can specify its own custom fields.

The first Adatum survey only has the **Product name** custom field in the **surveys** table; the second Adatum survey has both the **Product name** and **Owner** custom fields in the **surveys** table. The two Fabrikam surveys only have the **Promotion** custom field in the **surveys** table.

The Surveys application must be able to read from and write to the custom fields in the Surveys table. The developers at Tailspin considered two approaches to this. The first was to use a custom DLL for each tenant that was responsible for accessing the custom fields. The second approach was to store details of the custom fields as part of the tenant configuration data, and use this configuration data to determine how to access the custom fields at runtime.

Tailspin selected the second approach for two reasons: there is no need for Tailspin to create custom code for each tenant to support that tenant's custom fields. It is also easier to support different versions of the customization for a tenant. In the table of custom fields shown earlier, you can see that Adatum changes the set of custom fields it uses after it creates the first survey.

To read a more detailed discussion of **RowKeys** and **PartitionKeys** in Windows Azure table storage, see Chapter 7, *"Moving to Windows Azure Table Storage"* of the guide *"Moving Applications to the Cloud."*

Storing Tenant Data

The application collects most of the subscriber data during the on-boarding process. The **Logo** property of the **Tenant** class, shown below, contains the URL for the subscriber's logo. The application stores logo images in a public blob container named **logos**.

```C#
[Serializable]
public class Tenant
{
  public string ClaimType { get; set; }
  public string ClaimValue { get; set; }
  public string HostGeoLocation { get; set; }
  public string IssuerThumbPrint { get; set; }
  public string IssuerUrl { get; set; }
  public string IssuerIdentifier { get; set; }
  public string Logo { get; set; }
  [Required(ErrorMessage =
  "* You must provide a Name for the subscriber.")]
  public string Name { get; set; }
  public string WelcomeText { get; set; }
  public SubscriptionKind SubscriptionKind { get; set; }
  public Dictionary<string, List<UDFMetadata>>
          ExtensionDictionary { get; set; }
  public string SqlAzureConnectionString { get; set; }
  public string DatabaseName { get; set; }
  public string DatabaseUserName { get; set; }
  public string DatabasePassword { get; set; }
  public string SqlAzureFirewallIpStart { get; set; }
  public string SqlAzureFirewallIpEnd { get; set; }
}
```

We chose to store tenant data in Windows Azure blob storage to simplify the implementation of this part of the application. Tenant data has a very simple structure that can easily be stored in a blob, and it does not require any of the features provided by table storage. Tenant data is read very frequently and so, to minimize the number of storage transactions and to improve performance, the application caches tenant data in memory.

Storing Survey Answers

The Surveys application saves survey answers in blob storage. The application creates a blob container for each survey with a name that follows this pattern: **surveyanswers-<*tenant name*>-<*survey slug name*>**. This guarantees a unique container name for every survey and ensures that the application can easily identify the answers that belong to a specific survey or tenant.

> Tailspin chose to save each complete survey to blob storage rather than as a set of answer entities in a table because it found that saving to blob storage was faster in this particular scenario. The developers at Tailspin ran timing tests to compare saving to a blob and saving the same data as a set of rows in an Entity Group Transaction to table storage and found a significant difference between the two approaches. However, when using blob storage to persist survey responses Tailspin must also consider how subscribers can browse their survey responses in the order that they were submitted. For a more detailed explanation of how Tailspin evaluated the pros and cons of saving survey response data to blob or table storage, see Chapter 5, "Maximizing Availability, Scalability, and Elasticity."

For each completed survey response, the Surveys application saves a blob into the survey's container. The content of each blob is a **SurveyAnswer** object serialized in the JavaScript Object Notation (JSON) format. A **SurveyAnswer** object encapsulates a respondent's complete survey response, with all the respondent's answers to individual questions. The following code example shows the **SurveyAnswer** and **QuestionAnswer** classes. The **QuestionAnswers** property of the **SurveyAnswer** class is a list of **QuestionAnswer** objects.

```csharp
C#
public class SurveyAnswer
{
  ...

  public string SlugName { get; set; }
  public string Tenant { get; set; }
  public string Title { get; set; }
  public DateTime CreatedOn { get; set; }
  public IList<QuestionAnswer>
          QuestionAnswers { get; set; }
}

public class QuestionAnswer
{
  public string QuestionText { get; set; }
  public QuestionType QuestionType { get; set; }
  [Required(ErrorMessage = "* You must provide an answer.")]
  public string Answer { get; set; }
  public string PossibleAnswers { get; set; }
}
```

The name of the blob used to store each survey response is a GUID, which ensures that each blob in the container has a unique name. This means that the application does not save the survey responses with blob names that provide any useful ordering. Tailspin chose this approach in preference to saving the blobs using a naming convention that reflects the order in which the system received the survey answers in order to avoid the append/prepend anti pattern described in the presentation *Windows Azure Storage Deep Dive* on Channel 9.

However, Tailspin does have a requirement to enable subscribers to view the survey responses in the order they were submitted. To achieve this, the Surveys application also uses blob storage to store an ordered list of the responses to each survey. For each survey, the application stores a blob that contains a serialized List object containing the ordered names of all the survey response blobs (each of which contains a serialized **SurveyAnswer** object) for that survey. The **List** object is serialized in the JSON format. The section "Implementing Paging" later in this chapter explains how the Surveys application uses these List objects to enable paging through the survey results.

However, it's possible that a very large number of answers to a survey will affect performance because the process of updating this list will take longer as the size of the list grows. See the section "Maintaining a List of Survey Answers" in Chapter 7, "Managing and Monitoring Multi-tenant Applications," for more details of how Tailspin plans to resolve this issue.

Storing Survey Answer Summaries

The Surveys application uses blob storage to save the summary statistical data for each survey. For each survey, it creates a blob named **<tenant name>-<survey slug name>** in the **surveyanswers-summaries** container. The application serializes a **SurveyAnswersSummary** object in JSON format to save the data. A **SurveyAnswersSummary** object contains summary data for a survey, such as the total number of responses received that is stored in the **TotalAnswers** property. There is one **Survey-AnswersSummary** object for every survey.

The **QuestionAnswersSummaries** property of a **SurveyAnswersSummary** object contains a list of the questions in the survey. A **QuestionAnswerSummary** object contains the summary data for an individual question, such as an average for numeric questions. There is one **QuestionAnswerSummary** object for each question in a survey.

The following code example shows the **SurveyAnswersSummary** and **QuestionAnswersSummary** classes that define this summary data.

```C#
public class SurveyAnswersSummary
{

  ...

  public string Tenant { get; set; }
  public string SlugName { get; set; }
  public int TotalAnswers { get; set; }
  public IList<QuestionAnswersSummary>
    QuestionAnswersSummaries { get; set; }

  ...

}

public class QuestionAnswersSummary
{
  public string AnswersSummary { get; set; }
  public QuestionType QuestionType { get; set; }
  public string QuestionText { get; set; }
  public string PossibleAnswers { get; set; }
}
```

Notice that the summary is stored as a string for all question types, including numeric. This helps to minimize the number of changes that would be required to add a new question type to the Surveys application.

Comparing Paging Solutions

The developers at Tailspin considered two solutions to enable tenants to browse through survey responses, each based on a different storage model. The first option assumed that the application stored the survey response data in table storage. The second option, which was the one chosen, assumed that the application stored the survey response data in blob storage.

Paging with Table Storage

The developers at Tailspin looked at two features of the Windows Azure table storage API to help them design this solution. The first feature is the continuation token that you can request from a query, which enables you to execute a subsequent query that starts where the previous query finished. You can use a stack data structure to maintain a list of continuation tokens that you can use to go forward one page or back one page through the survey responses. You must then keep this stack of continuation tokens in the user's session state to enable navigation for the user.

> *For an example of this approach, see the section, "Implementing Paging with Windows Azure Table Storage" in Chapter 7, "Moving to Windows Azure Table Storage," of the guide "Moving Applications to the Cloud."*

The second useful API feature is the ability to run asynchronous queries against Windows Azure table storage. This can help avoid thread starvation in the web server's thread pool in the web role by offloading time-consuming tasks to a background thread.

Paging with Blob Storage

The assumption behind this solution is that each survey answer is stored in a separate blob. To access the blobs in a predefined order, you must maintain a list of all the blobs. You can then use this list to determine the identity of the previous and next blobs in the sequence and enable the user to navigate backward and forward through the survey responses. To support alternative orderings of the data, you must maintain additional lists.

Comparing the Solutions

Chapter 5, "Maximizing Availability, Scalability, and Elasticity," identifies cost savings as a reason to store survey responses directly in blob storage. In addition, paging with table storage is complex because you must manage the continuation stack in the user's session state.

What at first seems like the obvious solution (in this case, to use table storage) may not always turn out to be the best.

However, you must consider the costs and complexity associated with maintaining the ordered list of blobs in the second of the two alternative solutions. This incurs two additional storage transactions for every new survey response; one as the list is retrieved from blob storage, and one as it is saved back to blob storage. However, this still results in fewer transactions per survey response than the table-based solution. Furthermore, it's possible to avoid using any session state by embedding the links to the next and previous blobs directly in the web page.

The SQL Database Design

During the on-boarding process, the application will provision a new Windows Azure SQL Database instance for those subscribers who have access to this feature. This enables tenants to customize the database to their own requirements, and to manage their own custom reporting requirements using Windows Azure SQL Reporting. A private instance of Windows Azure SQL Database for each tenant also helps to ensure that survey response data remains confidential.

Giving every subscriber a separate instance of Windows Azure SQL Database allows the subscribers to customize the database to their own requirements. It also simplifies the security model, making it easier for Tailspin to ensure that survey response data is kept isolated.

The provisioning process will create the necessary tables in the database. As part of the on-boarding process, the Surveys application saves in blob storage (as part of the subscriber's details) the information that the application and the subscriber require to access the Windows Azure SQL Database instance.

A subscriber can use the UI to request the application export the survey data to a SQL Database instance. The UI notifies the worker role by placing a message on a queue. A task in a worker role monitors the queue for messages that instruct it to dump a subscriber's survey results to tables in Windows Azure SQL Database. Figure 3 shows the table structure in Windows Azure SQL Database.

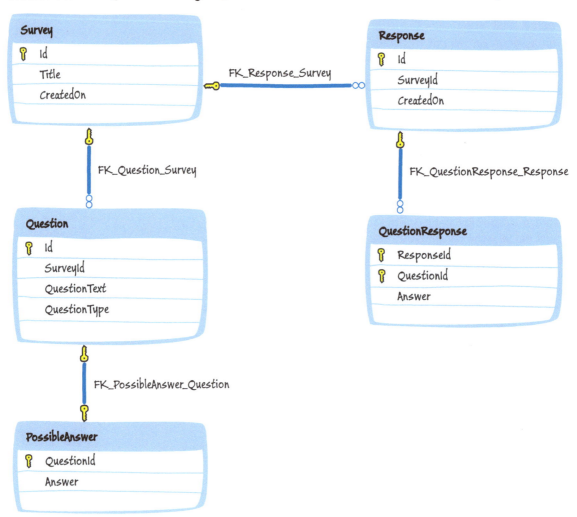

FIGURE 3
Surveys table structure in Windows Azure SQL Database

INSIDE THE IMPLEMENTATION

Now is a good time to walk through some of the code in the Tailspin Surveys application in more detail. As you go through this section, you may want to download the Visual Studio solution for the Tailspin Surveys application from *http://wag.codeplex.com/.*

The Data Store Classes

The Surveys application uses store classes to manage storage. This section briefly outlines the responsibilities of each of these store classes.

SurveyStore Class

This class is responsible for saving survey definitions to table storage and retrieving the definitions from table storage.

SurveyAnswerStore Class

This class is responsible for saving survey answers to blob storage and retrieving survey answers from blob storage. This class creates a new blob container when it saves the first response to a new survey, so there is one container per survey. It uses a queue to track new survey responses; the application uses this queue to calculate the summary statistical data for surveys.

This class also provides support for browsing sequentially through survey responses.

This class uses caching to reduce latency when retrieving survey definitions for the public web site.

SurveyAnswersSummaryStore Class

This class is responsible for saving summary statistical data for surveys to blobs in the surveyanswerssummaries container, and for retrieving this data.

SurveySqlStore Class

This class is responsible for saving survey response data to Windows Azure SQL Database. For more information, see the section "Implementing the Data Export" later in this chapter.

SurveyTransferStore Class

This class is responsible for placing a message on a queue when a subscriber requests the application to dump survey data to Windows Azure SQL Database.

TenantStore Class

This class is responsible for saving and retrieving subscriber data and saving uploaded logo images. In the sample code, this class generates some default data for the Adatum and Fabrikam subscribers. This class also uses caching to reduce latency when retrieving tenant information from blob storage.

Accessing Custom Data Associated with a Survey

Tenants with premium subscriptions can choose to define additional properties for their surveys. When a tenant creates a new survey, the UI allows that tenant to add user-defined fields and specify the values for the new survey. When the tenant views a list of surveys, the list includes the custom data values for each survey.

Defining a Tenant's Custom Fields

As part of the configuration data for premium tenants, Tailspin stores a dictionary of that tenant's custom fields. For example, if Adatum had chosen to use two custom fields, **ProductName** and **Owner** with its surveys, the blob that contains the Adatum configuration, would contain the following information:

```
JSON
"ExtensionDictionary":{"SurveyRow":[
  {
   "Name":"ProductName",
   "Type":5,
   "Display":"Product Name",
   "Mandatory":false,
   "DefaultValue":null
  },
  {
   "Name":"Owner",
   "Type":5,
   "Display":"Owner",
   "Mandatory":false,
   "DefaultValue":null
  }
]}
```

Every custom field has a name, a data type such as string or integer, a display value to use as a label in the UI, a Boolean flag to specify whether it is a required field, and an optional default value. A subscriber can add or delete custom field definitions on the **Model** extensions tab in the private tenant web site.

Writing Custom Fields to the Surveys Table

When a tenant creates a new survey, the private tenant web site UI reads the custom field definitions from the tenant configuration and adds appropriate UI elements to enable the tenant to add values to these fields. The Surveys application then persists these values in the custom fields to the surveys table when the tenant saves a survey.

The following code sample shows the definition of the **Survey** class that includes a list of user-defined fields. The **IUDFModel** interface defines the **UserDefinedFields** property.

```csharp
C#
[Serializable]
public class Survey : IUDFModel
{
  private readonly string slugName;

  ...

  public string Tenant { get; set; }

  [Required(ErrorMessage =
    "* You must provide a Title for the survey.")]
  public string Title { get; set; }

  public DateTime CreatedOn { get; set; }

  public List<Question> Questions { get; set; }

  public IList<UDFItem> UserDefinedFields { get; set; }
}
```

Figure 4 shows an overview of the way that the mechanism for saving a new survey includes the capability to save user-defined fields, and the core classes that are involved in the process. The following section of this chapter describes the process in more detail.

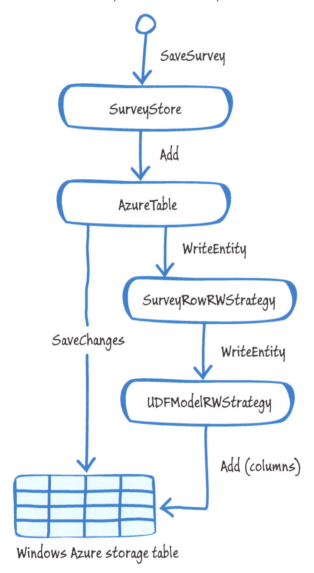

FIGURE 4
Overview of the mechanism for saving user-defined fields in a new survey

The **SaveSurvey** method in the **SurveyStore** class takes a **Survey** object as a parameter, and is responsible for creating a **SurveyRow** object and adding it as a new row to the Survey table. The **SurveyStore** class takes as parameters to its constructor instances of objects that implement the **IAzureTable** interface, one each for the Survey table and the Question table.

Objects that implement the **IAzureTable** interface are generic types that accept a table row type, such as **SurveyRow** and **QuestionRow**, and they expose a property named **ReadWriteStrategy** of type **IAzureTableRWStrategy** that is populated by a class that exposes the **ReadEntity** and **WriteEntity** methods. The following code sample from the **ContainerBootstraper** class in the Tailspin.Web project shows how the application registers the required types for dependency injection into the **SurveyStore** class, including an instance of the **SurveyRowRWStrategy** class for the **ReadWriteStrategy** property of the **AzureTable** class.

```C#
container.RegisterType<IUDFDictionary, UDFDictionary>();

container.RegisterType<IAzureTableRWStrategy,
    SurveyRowRWStrategy>(typeof(SurveyRow).Name);

var readWriteStrategyProperty = new InjectionProperty(
    "ReadWriteStrategy",
    new ResolvedParameter(
        typeof(IAzureTableRWStrategy),
        typeof(SurveyRow).Name));

container
  .RegisterType<IAzureTable<SurveyRow>,
    AzureTable<SurveyRow>>(
      new InjectionConstructor(cloudStorageAccountType,
            AzureConstants.Tables.Surveys),
      readWriteStrategyProperty,
      retryPolicyFactoryProperty)
...
```

The **SaveSurvey** method of the **SurveyStore** class saves a **SurveyRow** instance to table storage by calling the **Add** method of the **AzureTable** instance. This method creates a new **TableServiceContext** to use to access the Windows Azure table by calling the **CreateContext** method defined in the **AzureTable** class. The **CreateContext** method hooks up the **ReadEntity** and **WriteEntity** methods of the **SurveyRowRWStrategy** class to the **ReadingEntity** and **WritingEntity** events of the **Table-ServiceContext**, as shown in the following code sample.

```csharp
C#
private TableServiceContext CreateContext()
{
  ...

  if (this.ReadWriteStrategy != null)
  {
    context.ReadingEntity += (sender, args)
      => this.ReadWriteStrategy.ReadEntity(context, args);
    context.WritingEntity += (sender, args)
      => this.ReadWriteStrategy.WriteEntity(context, args);
  }

  return context;
}
```

The **SurveyRowRWStrategy** class inherits from the **UDFModelRWStrategy** class and does not override the **WriteEntity** method. The following code sample shows the **WriteEntity** method from the **UDFModelRWStrategy** class, which creates the user-defined fields and adds them to the table schema.

```csharp
C#
public virtual void WriteEntity(
      TableServiceContext context,
      ReadingWritingEntityEventArgs args)
{
  var ns = XNamespace.Get(DATASERVICESNS);
  var nsmd = XNamespace.Get(DATASERVICESMETADATANS);
  var survey = args.Entity as SurveyRow;
  if (survey != null && survey.UserDefinedFields != null
      && survey.UserDefinedFields.Count > 0)
  {
    var properties = args.Data
        .Descendants(nsmd + "properties").First();
    foreach (var udfItem in survey.UserDefinedFields)
    {
      var udfField = new XElement(ns + udfItem.Name,
                                  udfItem.Value);
      udfField.Add(new XAttribute(nsmd + "type",
                                  udfItem.GetEdmType()));
      properties.Add(udfField);
    }
  }
}
```

For more information about this method of saving entities to Azure table storage, see the blog post *"Entities in Azure Tables."*

Reading Custom Fields from the Surveys Table

Saving a survey definition to Windows Azure table storage always includes the custom fields that are currently defined in the tenant's configuration data. When a survey definition is read it must include the custom fields that were defined when the survey was originally saved.

Figure 5 shows an overview of the way that the mechanism for reading a survey definition includes the capability to retrieve the user-defined fields and add them to the **SurveyRow** that is returned, and the core classes that are involved in the process.

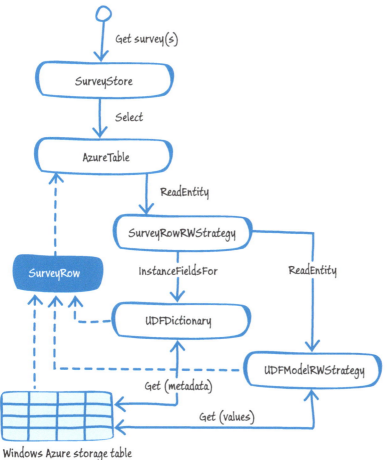

FIGURE 5
Overview of the mechanism for retrieving user-defined fields for a survey

The **SurveyStore** class executes a **Select** query against the **AzureTable** instance that was injected into it when it was instantiated. The **ReadingEntity** event that occurs when reading entities from the **TableServiceContext** causes the **ReadEntity** method in the **SurveyRowRWStrategy** class to execute.

The **SurveyRowRWStrategy** class holds a reference to an instance of a class that implements the **IUDFDictionary** interface. The dependency injection registration you saw earlier causes this to be populated with an instance of the **UDFDictionary** class.

The **ReadEntity** method of the **SurveyRowRWStrategy** class calls the **InstanceFieldsFor** method of the **UDFDictionary** class to discover the names of the fields from the table metadata, and then calls the **ReadEntity** method of the **UDFModelRWStrategy** class to get the field values from the table itself. The **SurveyRowRWStrategy** class then assigns the collection of user-defined fields to the **UserDefinedFields** property of the **SurveyRow** instance.

Implementing Paging

The code walkthrough in this section is divided into two parts. The first describes how the application maintains an ordered list of blobs. The second describes how the application uses this list to page through the responses.

Maintaining the Ordered List of Survey Responses

Tailspin Surveys saves and processes survey response data using two asynchronous tasks hosted in a worker role. For more information about the implementation that Tailspin chose for saving and processing survey response data, see Chapter 5, "Maximizing Availability, Scalability, and Elasticity." This section focuses on how the data architecture in Tailspin Surveys supports paging through Survey responses stored in blobs.

The **PostRun** method in the **UpdatingSurveyResultsSummaryCommand** class in the worker role calls the **AppendSurveyAnswerIdsToAnswerList** method for the collection of new survey responses that the task processed in the **Run** method. The following code example shows how the **AppendSurvey-AnswerIdsToAnswerList** method in the **SurveyAnswerStore** class retrieves the list of survey responses from a blob, adds the new survey responses to the list, and saves the list back to blob storage.

```C#
public void AppendSurveyAnswerIdsToAnswersList(
                string tenant, string slugName,
                IEnumerable<string> surveyAnswerIds)
{
  OptimisticConcurrencyContext context;
  string id = string.Format(CultureInfo.InvariantCulture,
                            "{0}-{1}", tenant, slugName);
  var answerIdList = this.surveyAnswerIdsListContainer
           .Get(id, out context) ?? new List<string>(1);
  answerIdList.AddRange(surveyAnswerIds);
  this.surveyAnswerIdsListContainer
           .Save(context, answerIdList);
}
```

The application stores the list of survey responses in a List object, which it serializes in the JSON format and stores in a blob. There is one blob for every survey and the application stores all of these blobs in the same blob container.

For more information about the concurrency management that the Surveys application uses when it saves the list of survey answers, see the section "Pessimistic and Optimistic Concurrency Handling" in Chapter 5, "Maximizing Availability, Scalability, and Elasticity."

Implementing the Paging

When the Surveys application displays a survey response, it finds the blob that contains the survey response by using a blob ID. It can use the ordered list of blob IDs to create navigation links to the next and previous survey responses.

The following code example shows the **BrowseResponses** action method in the **SurveysController** class in the TailSpin.Web project.

The application adds new responses to the queue in the order that they are received. When it retrieves messages from the queue and adds the blob IDs to the list, it preserves the original ordering.

```C#
public ActionResult BrowseResponses(string tenant,
    string surveySlug, string answerId)
{
  SurveyAnswer surveyAnswer = null;
  if (string.IsNullOrEmpty(answerId))
  {
    answerId = this.surveyAnswerStore
        .GetFirstSurveyAnswerId(tenant, surveySlug);
  }

  if (!string.IsNullOrEmpty(answerId))
  {
    surveyAnswer = this.surveyAnswerStore
        .GetSurveyAnswer(tenant, surveySlug, answerId);
  }

  var surveyAnswerBrowsingContext = this.surveyAnswerStore
      .GetSurveyAnswerBrowsingContext(tenant,
          surveySlug, answerId);

  var browseResponsesModel = new BrowseResponseModel
    {
      SurveyAnswer = surveyAnswer,
      PreviousAnswerId =
          surveyAnswerBrowsingContext.PreviousId,
      NextAnswerId = surveyAnswerBrowsingContext.NextId
    };
```

```
      var model = this.CreateTenantPageViewData
                        (browseResponsesModel);
    model.Title = surveySlug;
    return this.View(model);
}
```

This action method uses the **GetSurveyAnswer** method in **Survey-AnswerStore** class to retrieve the survey response from blob storage and the **GetSurveyAnswerBrowsingContext** method to retrieve a **SurveyBrowsingContext** object that contains the blob IDs of the next and previous blobs in the sequence. It then populates a model object with this data to forward on to the view.

Implementing the Data Export

The following code example shows the task in the worker role that executes when it is triggered by a message in a queue. Chapter 4, "Partitioning Multi-Tenant Applications," describes the message queues and the worker role in Tailspin Surveys in more detail. This section focuses on how the application handles data and the data export process.

You can find the **Run** method that performs the data export in the **TransferSurveysToSqlAzureCommand** class in the worker role project. The **SurveyTransferMessage** class identifies the tenant who owns the data, and the survey to export.

This task is part of the worker role described in Chapter 4, "Partitioning Multi-Tenant Applications." A message in a queue triggers this task.

```
C#
public bool Run(SurveyTransferMessage message)
{
  Tenant tenant =
        this.tenantStore.GetTenant(message.Tenant);
  this.surveySqlStore.Reset(
        tenant.SqlAzureConnectionString, message.Tenant,
        message.SlugName);

  Survey surveyWithQuestions = this.surveyStore
        .GetSurveyByTenantAndSlugName(message.Tenant,
            message.SlugName, true);

  IEnumerable<string> answerIds = this.surveyAnswerStore
        .GetSurveyAnswerIds(message.Tenant,
            surveyWithQuestions.SlugName);

  SurveyData surveyData = surveyWithQuestions.ToDataModel();

  foreach (var answerId in answerIds)
```

```
{
  SurveyAnswer surveyAnswer = this.surveyAnswerStore
      .GetSurveyAnswer(surveyWithQuestions.Tenant,
          surveyWithQuestions.SlugName, answerId);

  var responseData = new ResponseData
    {
      Id = Guid.NewGuid().ToString(),
      CreatedOn = surveyAnswer.CreatedOn
    };

  foreach (var answer in surveyAnswer.QuestionAnswers)
  {
    QuestionAnswer answerCopy = answer;
    var questionResponseData = new QuestionResponseData
      {
        QuestionId = (
          from question in surveyData.QuestionDatas
          where question.QuestionText ==
                answerCopy.QuestionText
          select question.Id).FirstOrDefault(),
        Answer = answer.Answer
      };
    responseData.QuestionResponseDatas
            .Add(questionResponseData);
  }
  if (responseData.QuestionResponseDatas.Count > 0)
  {
    surveyData.ResponseDatas.Add(responseData);
  }
}

this.surveySqlStore
    .SaveSurvey(tenant.SqlAzureConnectionString,
                surveyData);
return true;
}
```

The method first resets the survey data in Windows Azure SQL Database before it iterates over all the responses to the survey and saves the data to the tenant's SQL Database instance. The application does not attempt to parallelize this operation; for subscribers who have large volumes of data, the dump operation may run for some time.

The application uses LINQ to SQL to manage the interaction with Windows Azure SQL Database. The following code from the **SurveySqlStore** class shows how the application uses the **SurveyData** and **SurveySqlDataContext** classes. The SurveySql.dbml designer creates these classes.

```csharp
C#
public void SaveSurvey(string connectionString,
  SurveyData surveyData)
{
  this.ConnectionRetryPolicy.ExecuteAction(() =>
  {
    using (var dataContext =
      new SurveySqlDataContext(connectionString))
    {
      dataContext.SurveyDatas.InsertOnSubmit(surveyData);
      try
      {
        this.CommandRetryPolicy.ExecuteAction(
          () => dataContext.SubmitChanges());
      }
      catch (SqlException ex)
      {
        TraceHelper.TraceError(ex.TraceInformation());
        throw;
      }
    }
  });
}
```

This code uses the Transient Fault Handling Application Block to handle any transient errors when it tries to save the data to SQL Database.

Displaying Questions

The application stores the definition of a survey and its questions in table storage. To render the questions in a page in the browser from the Tailspin.Web.Survey.Public web application project, the application uses the MVC **EditorExtensions** class.

Tailspin chose this mechanism to render the questions because it makes it easier to extend the application to support additional question types.

When the **Display** action method in the **SurveysController** class in the TailSpin.Web.Survey.Public project builds the view to display the survey, it retrieves the survey definition from table storage, populates a model, and passes the model to the view. The following code example shows this action method.

```C#
[HttpGet]
public ActionResult Display(string tenant,
                            string surveySlug)
{
  var surveyAnswer = CallGetSurveyAndCreateSurveyAnswer(
      this.surveyStore, tenant, surveySlug);

  var model =
      new TenantPageViewData<SurveyAnswer>(surveyAnswer);
  if (surveyAnswer != null)
  {
    model.Title = surveyAnswer.Title;
  }
  return this.View(model);
}
```

The view uses the **EditorExtensions** class to render the questions. The following code example shows how the Display.aspx page uses the **Html.EditorFor** element that is defined in the **System.Web.Mvc. EditorExtensions** class.

```HTML
<% for (int i = 0;
    i < this.Model.ContentModel.QuestionAnswers.Count; i++)
{ %>
  ...
<%: Html.EditorFor(m=>m.ContentModel.QuestionAnswers[i],
      QuestionTemplateFactory.Create(
        Model.ContentModel.QuestionAnswers[i].QuestionType)) %>
  ...
<% } %>
```

This element iterates over all the questions that the controller retrieved from storage and uses the **QuestionTemplateFactory** utility class to determine which user control (an .ascx file) to use to render the question. The user controls FiveStar.ascx, MultipleChoice.ascx, and SimpleText.ascx are in the EditorTemplates folder in the project.

Displaying the Summary Statistics

The asynchronous task that generates the summary statistics from survey responses (this task is described in Chapter 5, "Maximizing Availability, Scalability, and Elasticity") stores the summaries in blob storage. It uses a separate blob for each survey. The Surveys application displays these summary statistics in the same way that it displays questions. The following code example shows the **Analyze** action method in the **SurveysController** class in the TailSpin.Web project that reads the results from blob storage and populates a model.

```csharp
C#
public ActionResult Analyze(string tenant,
                            string surveySlug)
{
  var surveyAnswersSummary =
      this.surveyAnswersSummaryStore
        .GetSurveyAnswersSummary(tenant, surveySlug);

  var model =
      this.CreateTenantPageViewData(surveyAnswersSummary);
  model.Title = surveySlug;
  return this.View(model);
}
```

The view uses the **Html.DisplayFor** element to render the questions. The following code example shows a part of the Analyze.aspx file.

```html
HTML
<% for (int i = 0;
        i < this.Model.ContentModel
              .QuestionAnswersSummaries.Count; i++)
{ %>
<li>
<%: Html.DisplayFor(m =>
      m.ContentModel.QuestionAnswersSummaries[i],
      "Summary-" + TailSpin.Web.Survey.Public.Utility
        .QuestionTemplateFactory.Create
          (Model.ContentModel.QuestionAnswersSummaries[i]
            .QuestionType))%>
</li>
<% } %>
```

The user control templates for rendering the summary statistics are named Summary-FiveStar.ascx (which displays an average for numeric range questions), Summary-MultipleChoice.ascx (which displays a histogram), and Summary-SimpleText.ascx (which displays a word cloud). You can find these templates in the DisplayTemplates folder in the TailSpin.Web project. To support additional question types you must add additional user control templates to this folder.

MORE INFORMATION

All links in this book are accessible from the book's online bibliography available at: *http://msdn.microsoft.com/library/jj871057.aspx.*

For more information about Windows Azure storage services, see *"Data Services"* and the *Windows Azure Storage Team Blog.*

For a comparison of Windows Azure Table Storage and Windows Azure SQL Database, see *"Windows Azure Table Storage and Windows Azure SQL Database - Compared and Contrasted."*

For further information about continuation tokens and Windows Azure table storage, see the section, "Implementing Paging with Windows Azure Table Storage" in Chapter 7, *"Moving to Windows Azure Table Storage"* of the guide *"Moving Applications to the Cloud."*

4

Partitioning Multi-Tenant Applications

This chapter examines architectural and implementation considerations in the Surveys application from the perspective of building a multi-tenant application. Questions such as how to partition the application, how users will interact with the application, how you can configure it for multiple tenants, and how you handle session state and caching are all directly relevant to a multi-tenant architecture. This chapter describes how Tailspin resolved these questions for the Surveys application. For other applications, different choices may be appropriate.

PARTITIONING A WINDOWS AZURE APPLICATION

There are two related reasons for partitioning a Windows Azure application. First, physically partitioning the application enables you to scale the application out. For example, by running multiple instances of the worker roles in the application you can achieve higher throughput and faster processing times. Second, in a multi-tenant application, you either logically or physically partition the application to provide isolation between the tenants. Isolating tenants' data helps to ensure that data belonging to a tenant is kept private but also helps to manage the application. For example, you may want to provide different levels of service to different tenants by scaling them out differently, or provide a different set of features to some tenants based on the type of subscription they have.

A Windows Azure application is typically comprised of multiple elements such as worker roles, web roles, queues, storage, and caching. If your application is a multi-tenant application, you can chose between different models for each of these elements:

- **Single instance, multi-tenant model**. For example, a single instance of a queue handles messages for all the tenants in your application.
- **Multi-instance, single-tenant model**. For example, each tenant has its own private message queue.
- **Multi-instance, multi-tenant model**. For example, premium tenants each have their own queues, but standard tenants use a shared queue.

Chapter 2, "Hosting a Multi-Tenant Application on Windows Azure," describes the differences between these models in more detail, along with a discussion of criteria you should consider when you are choosing between them for any particular part of the application. Chapter 3, "Choosing a Multi-Tenant Data Architecture," addresses these issues in relation to data storage in a Windows Azure application. This chapter focuses on partitioning web and worker roles, queues, and caches. It examines this partitioning from the perspective of the choices that Tailspin made in the design and implementation of the Surveys application.

Figure 1 illustrates the relationships between the various Windows Azure elements discussed in this chapter.

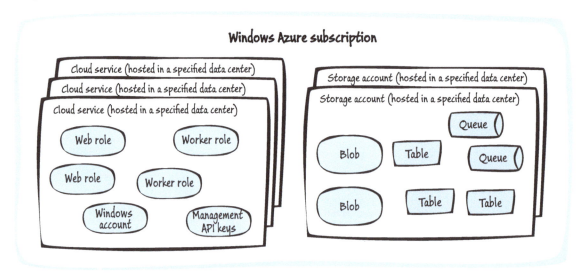

FIGURE 1
Key Windows Azure elements discussed in this chapter

Some key points to note from Figure 1 are:

- Administrative access is at the level of a Windows Azure subscription using a Windows Account or a Management API key.
- A Windows Azure subscription can contain multiple cloud services and multiple storage accounts.
- Every cloud service is assigned a unique DNS name.
- Each cloud service and each storage account is hosted in a data center selected by the subscription administrator.

Partitioning Web and Worker Roles

A Windows Azure application is typically comprised of multiple role types. For example, the Tailspin Surveys application has two web roles (one for the public website and one for the private tenant website), and a single worker role. When you deploy a web or worker role to Windows Azure, you deploy it to a cloud service that itself is part of a Windows Azure subscription. The options available for partitioning a deployment by tenant are described in the following table:

Partitioning scheme	Notes
One subscription per tenant	Makes it easy to bill individual tenants for the compute resources they consume.
	Enables tenants to provide their own compute resources and then manage them.
	During the provisioning process, the tenant would need to provide access details such as Management API keys or Microsoft account credentials if you are going to deploy the application to the tenant's subscription.
	You need to be careful about the location of the cloud service that hosts the roles in relation to the location of the cloud storage the application uses in order to control data transfer costs and minimize latency.
	Provisioning a new Windows Azure subscription is a manual process.
	This does not allow tenants to share resources and costs. Each tenant must pay for all the role instances it uses.
	This scheme also implies one tenant per cloud service.
Group multiple tenants in a subscription	If your tenants can subscribe to different levels of functionality (such as basic, standard, and premium) for the application, then using different subscriptions makes it easier to track the costs of providing each level of functionality.
	You must still partition the application for different tenants within a subscription using one of the other partitioning schemes.
One tenant per cloud service	Cloud services can be provisioned automatically.
	Each tenant can run the application in a geographic region of their choice.
	Makes it easy to assign costs to tenants because each cloud service is a separate line item on the Windows Azure bill.
	Provides for a very high degree of tenant isolation.
	This does not allow tenants to share resources and costs. Each tenant must pay for all the role instances it uses.
Group multiple tenants in a cloud service	You can use cloud services to group tenants based on geographic region or on different levels of functionality.
	You must still partition the application for tenants within a cloud service using one of the other partitioning schemes, such as using multi-tenant roles.
	Tenants in a cloud service share the resources and costs associated with that service.
Group multiple tenants in a role	Requires the web or worker role to support multi-tenancy.
	Tenants share the resources and costs associated with the role.

*It is your responsibility to
ensure that your web roles
can identify the tenant in the
request, and to ensure that web
roles preserve the isolation
between your tenants.*

At the time of writing, there is a soft limit of 20 cores per Windows Azure subscription. With this limit, you could deploy 20 small role instances, 10 medium role instances, or 5 large role instances. This limit makes any solution that assigns tenants to roles in a one-to-one relationship within a subscription impractical for any multi-tenant application with more than a small number of tenants.

Identifying the Tenant in a Web Role

Every cloud service must have a unique DNS name. Therefore, if you have one tenant per cloud service, each tenant can use a unique DNS name to access its copy of the application.

However, if you have multiple tenants sharing web roles within a cloud service, you must have some way to identify the tenant for each web request that accesses tenant specific data. Once you know which tenant the web request is associated with you can ensure that any queries or data updates operate only on the data or other resources that belong to that tenant.

There are a number of options for identifying the tenant from a web request.

- **Authentication.** If users accessing the site must authenticate, the site can determine the tenant from the authenticated identity.
- **The URL path.** For example, Tailspin's tenants Adatum and Fabrikam could use *http://tailspinsurveys.cloudapp.net/adatum/ surveys* and *http://tailspinsurveys.cloudapp.net/fabrikam/surveys*.
- **The subdomain.** For example, Tailspin's tenants Adatum and Fabrikam could use *http://adatum.tailspinsurveys.com* and *http:// fabrikam.tailspinsurveys.com*.
- **A custom domain.** For example, Tailspin's tenants Adatum and Fabrikam could use *http://surveys.adatum.com* and *http://surveys. fabrikam.com*.

Option 1 — Using Authentication

For the first option, using authentication, you can use any standard authentication mechanism as long as your application can determine from the authenticated identity the tenant that should be associated with the request. You don't need to include the tenant ID anywhere in the URL, so you can use the same domain and path for all tenant requests for a specific service. For example, Tailspin could use *http:// tailspinsurveys.cloudapp.net/surveys* to enable a tenant to access a list of all its surveys.

Tailspin could also add a CNAME entry to its DNS configuration to map a custom subdomain to the surveys application. For example, Tailspin could map *http://surveys.tailspin.com* to *http://tailspinsurveys.cloudapp.net/surveys*.

This approach also enables you to protect your site using SSL by uploading to your cloud service a standard SSL certificate that identifies a single domain, and then configuring an HTTPS endpoint that uses this certificate.

Option 2 — Using the URL Path

If you are using ASP.NET MVC it is very easy to use MVC routing to identify the tenant from an element in the path. This approach is useful if you don't want to use authentication (for example, on a public site) but you do need to identify the tenant.

You can use the same domain for all requests. For example, Tailspin could use *http://tailspinsurveys.cloudapp.net/{tenant}/surveys*, where *{tenant}* is the ID of a tenant, to enable public access to a list of tenant surveys.

This approach also enables you to protect your site using SSL by uploading to your cloud service a standard SSL certificate that identifies a single domain, and then configuring an HTTPS endpoint that uses this certificate.

Option 3 — Using a Subdomain for Each Tenant

Although routing based on a subdomain is not part of the standard MVC routing functionality, it is possible to implement this behavior in MVC. For an example, see the blog post *"ASP.NET MVC Domain Routing."* This approach is also useful if you don't want to use authentication (for example, on a public site) but you do need to identify the tenant.

To be able to use a separate subdomain for each tenant you must create a CNAME entry for each tenant in your DNS configuration. For example, Tailspin's DNS configuration might look like this:

adatum.tailspinsurveys.com CNAME tailspinsurveys.cloudapp.net

fabrikam.tailspinsurveys.com CNAME tailspinsurveys.cloupapp.net

Some DNS providers enable you to use a wildcard CNAME entry so that you don't need to create an entry for every tenant. For example:

*.tailspinsurveys.com CNAME tailspinsurveys.cloudapp.net

Option 4 — Enabling Tenants to Use Custom Domains

There are several possible approaches to enable each tenant to map a custom domain to your Windows Azure application.

You can use host headers to map each tenant to a separate website in the cloud service. Although this enables you to have a separate website for every tenant, you must redeploy the application whenever you need to configure a new tenant because this approach requires entries in the Windows Azure service definition file.

> For more information about how to configure host headers in a Windows Azure application, see *"How to Configure a Web Role for Multiple Web Sites,"* and the useful walkthrough *"Exercise 1 Registering Sites, Applications, and Virtual Directories"* in the Windows Azure Training Course.

As an alternative you could allow each tenant to create its own DNS entry that maps a domain or subdomain owned by the tenant to one of your application's DNS names. For example, Adatum could create the following DNS entry in its DNS configuration:

surveys.adatum.com CNAME adatum.tailspinsurveys.com

Or Adatum could create the following entry:

surveys.adatum.com CNAME www.tailspinsurveys.net

In either case you could use the custom domain name in your ASP. NET routing by using the technique suggested in option 3 above, or use the **Request.Url** class in your code to identify the domain.

Using SSL with Windows Azure Cloud Services

Windows Azure uses the name of your cloud service to generate a unique subdomain at cloudapp.net for every cloud service; for example, tailspinsurveys.cloudapp.net. You can configure your DNS provider to point one or more of your subdomains to your cloudapp. net subdomain. For example, Tailspin could configure the following CNAME entries:

adatum.tailspinsurveys.com CNAME tailspinsurveys. cloudapp.net

fabrikam.tailspinsurveys.com CNAME tailspinsurveys. cloupapp.net

admin.tailspinsurveys.com CNAME tailspinsurveys.cloup-app.net

Each Windows Azure cloud service can only have a single SSL certificate. Typically, an SSL certificate is valid for a single subdomain, so Tailspin might choose to upload an SSL certificate that is valid for *admin.tailspinsurveys.com*. The other two subdomains would only be able to use the HTTP protocol. However, it is possible to purchase a wildcard SSL certificate. If Tailspin purchased an SSL certificate that is valid for **.tailspinsurveys.com*, then all *tailspinsurveys* subdomains could use the HTTPS protocol.

> *You can have multiple web roles within a Windows Azure cloud service, but each web role must listen on a different port. Because Internet Information Services (IIS) can have only one SSL certificate associated with a port, a multi-tenant application can use only a single domain name in a cloud service on the default HTTPS port 443.*

If you allow tenants to map their own domain or subdomain to your Windows Azure cloud application, you should verify that the tenant owns the domain before you customize your application to recognize the tenant's domain. Also, if you allow tenants to map their own DNS entries to your Windows Azure application, they should map to one of your DNS names instead of the underlying cloudapp. net address. This gives you the ability to manage redirections if you need to temporarily point your tenants to a different cloud service. This may be useful during updates of the application, or if a Windows Azure data center becomes temporarily unavailable.

Identifying the Tenant in a Worker Role

In a multi-tenant Windows Azure application, tenants typically share worker role instances. Either all tenants share a worker role; or groups of tenants share each worker role. Figure 2 shows three possible models.

Model 1

Tailspin cloud service

Tailspin worker role

Model 2

Tailspin cloud service

Tailspin standard worker role

Tailspin premium worker role

Model 3

Tailspin U.S. cloud service

Tailspin worker role

Tailspin European cloud service

Tailspin worker role

Figure 2
Different models for multi-tenant worker roles

In Model 1 all the tenants share all the worker role instances. In Model 2 all standard tenants use one set of worker role instances, and all premium tenants use another set; this enables you to scale the worker roles independently or provide different features for the different groups of tenants. Model 3 enables you to deploy the same worker role into different geographic regions; Tailspin plans to use this model.

Tailspin plans deploy the same worker role into different geographic regions, but also plans to prioritize the work for premium subscribers within the role.

Whichever model you choose, worker roles typically receive messages from queues, and these messages cause the worker role to perform some work on behalf of a tenant. In a multi-tenant application, you must be able to identify the tenant associated with the message that the worker role receives. There are two ways to identify the tenant: either every message contains the tenant ID, or every tenant has its own queue.

In some scenarios, you may also need to identify the type of tenant, such as whether the tenant has a standard or a premium subscription. If you are processing work for both types of tenant in the same role instances then, again, you can either include the tenant type in every message or use different message queues for each type of tenant. If you want to give priority to messages from tenants with premium subscriptions, using different queues for the different tenant types makes this easy to accomplish.

Partitioning Queues

You create Windows Azure queues within a Windows Azure storage account. By default you are limited to five storage accounts per subscription but you can request additional accounts by contacting Windows Azure customer support. Although you could use different storage accounts for different tenants, this approach is typically useful only if each tenant uses its own Windows Azure subscription for its Windows Azure queues.

Partitioning queues by using different subscriptions can be useful if you have a small number of tenants with very high throughput.

The previous discussion of worker role partitioning highlighted the use of queues as the way that web roles typically pass information to worker roles. You can use queues to partition the messages for tenants in three ways, as shown in Figure 3.

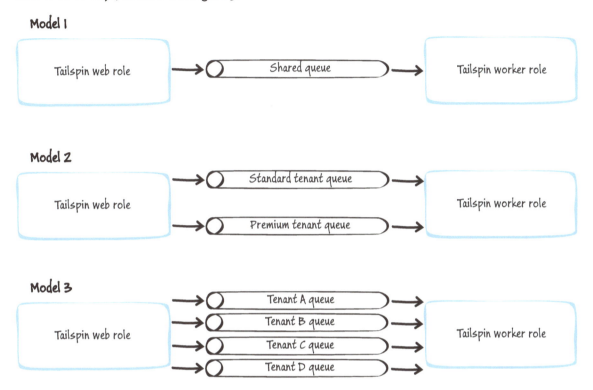

FIGURE 3
Different models for partitioning queues

The first model is useful if you do not need to distinguish between types of tenant. The second model makes it easy for the worker role to identify and process messages from different groups of tenants; Tailspin uses this approach to enable the worker role to prioritize messages from tenants with premium subscriptions. The third model is useful if you have very high volumes of messages or need the ability to manage each tenant's queue individually.

Within a storage account you can create as many queues as you need. You are charged based on the number of messages that you send, so there is no financial penalty in using multiple queues. However there are some limits on the total number of transactions per second and the bandwidth for each storage account. For more information, see *Best Practices for the Design of Large-Scale Services on Windows Azure Cloud Services* on MSDN.

Using a separate queue for each tenant has some potential scaling issues. For example, if you have 2,000 tenants, the consumer in the worker role must loop through and monitor 2,000 queues looking for work.

It is your responsibility to ensure that the application uses the correct queue for any message associated with a specific client. Windows Azure does not provide any configuration options that enable you to set permissions that limit access to a queue to a specific tenant or message type.

> *You can also use Windows Azure Service Bus queues to transport messages from web roles to worker roles. For more information about the differences between the two types of queues, see "Windows Azure Queues and Windows Azure Service Bus Queues - Compared and Contrasted."*

Partitioning Caches

At the time of writing, Windows Azure offers two caching models: Windows Azure Caching and Shared Caching. Windows Azure Caching uses one or more of the roles in your application to host cached data, whereas Shared Caching is a separate service that hosts cached data outside of your application.

When you configure Windows Azure Caching it creates a cache that is private to your application. However all of the tenants who use the application will, by default, have access to all of the data in the cache, so it is your responsibility to ensure that the design of your application prevents tenants from accessing cached data that belongs to other tenants.

It is possible to create multiple named caches in your application either by adding dedicated caching roles or by configuring caching on multiple roles. When you read or write to a Windows Azure Caching cache, you can specify the named cache you want to use.

You can also subdivide a named Windows Azure Caching cache into named regions. Regions are useful for grouping cached items together and enable you to tag items within a region. You can then query for items within a region by using tag values. You can also remove all the items cached in a single region with a single API call.

Although you could use different named caches to enable single-tenant caches, creating a new named cache requires changing your service configuration. Typically, you use multiple, named caches where you need multiple caching policies. For example, Tailspin could decide to use different caching policies for premium and standard tenants where the time to live for sessions belonging to premium subscribers is set to 30 minutes, but for standard subscribers is set to only five minutes.

Windows Azure Shared Caching enables you to create separate named caches, each of which has a maximum size. However, because you are charged for each shared cache that you create, this would be an expensive solution if you wanted to have one cache per tenant.

GOALS AND REQUIREMENTS

This section describes the goals and requirements Tailspin has for the Surveys application that relate to partitioning it as a multi-tenant application.

Isolation

When subscribers to the Tailspin Surveys application access their subscription details, survey definitions, and survey results, they should only see their own data. Subscribers must authenticate with the application to gain access to this private data.

It is a key requirement of the application to protect survey designs and results from unauthorized access, and the application will use a claims-based infrastructure to achieve this. Chapter 6, "Securing Multi-Tenant Applications," discusses the authentication and authorization mechanisms in the Tailspin Surveys application.

When survey respondents visit the Tailspin Surveys public site to complete a survey, they don't need to authenticate. However, survey respondents should not have access to survey response data, and the response data must be private to the subscriber who published the survey.

Scalability

The Tailspin Surveys application must be scalable. The partitioning schemes that Tailspin uses for the web and worker roles and for the Windows Azure queues should not limit Tailspin's ability to scale out the application. The web roles, the worker role, and the Windows Azure queues must be capable of being scaled independently of each other.

Three distinct groups of users will access the Surveys application: administrators at Tailspin who will manage the application, subscribers who will be creating their own custom surveys and analyzing the results, and users who will be filling out their survey responses. The first two groups will account for a very small proportion of the total number of users at any given time; the vast majority of users will be people who are filling out surveys. A large survey could have hundreds of thousands of users filling it out, while a subscriber might create a new survey only every couple of weeks.

Although you need to authenticate using an ACS key when you use a Windows Azure shared cache, this identity is only used to access the shared cache and cannot be used to control access to individual items in the cache.

There are three distinct groups of users who will use the Surveys application.

Furthermore, the numbers of users filling out surveys will be subject to sudden, large, short-lived spikes as subscribers launch new surveys. In addition to the different scalability requirements that arise from the two usage profiles, other requirements such as security will also vary.

> *Chapter 5, "Maximizing Availability, Scalability, and Elasticity," covers issues around scaling the Tailspin Surveys application in greater depth.*

Accessing the Surveys Application

There will be a single URL for the subscriber website where subscribers will need to authenticate before accessing their query designs and survey results data. Additionally, all access to the application by subscribers and administrators will use HTTPS to protect the data transferred between the application and the client.

The public website where users complete surveys will not require authentication. Survey respondents should be given some indication of the identity of the survey author through the URL they use to access the survey questions, and optionally through branding applied to the web pages. In the future, Tailspin also plans to enable subscribers to include a public landing page accessible from a URL that includes the subscriber's name, and which lists all of the surveys published by that subscriber.

Public surveys do not require HTTPS. This enables the use of DNS CNAME entries that define custom URLs for users to access and fill out these surveys.

Subscribers and survey respondents may be in different geographical locations. For example, a subscriber may be based in the U.S. but wants to perform some market research in Europe. Tailspin can minimize the latency for survey respondents by enabling subscribers to host their surveys in a datacenter located in an appropriate geographical region. However, subscribers may need to analyze the results collected from these surveys in their own geographical location.

Premium Subscriptions

Tailspin plans to offer multiple levels of subscription, initially standard and premium, with the ability to add further levels in the future. Tailspin wants to be able to offer different functionality and different levels of service with the different subscriptions. Initially, Tailspin will give priority to premium tenants ensuring that the worker role processes and saves their data faster than for standard tenants.

Windows Azure enables you to deploy role instances to datacenters in different geographic locations. Tailspin can host the subscriber and the survey web roles in different datacenters and use Windows Azure Traffic Manager to automatically route requests to the most appropriate datacenter. For more information see Chapter 5, "Maximizing Availability, Scalability, and Elasticity."

Designing Surveys

When a user designs a new survey in the Surveys application, they create the survey and then add questions one-by-one to the survey until it's complete. Figure 4 shows the sequence of screens, from an early mockup of this part of the UI, when a user creates a survey with two questions.

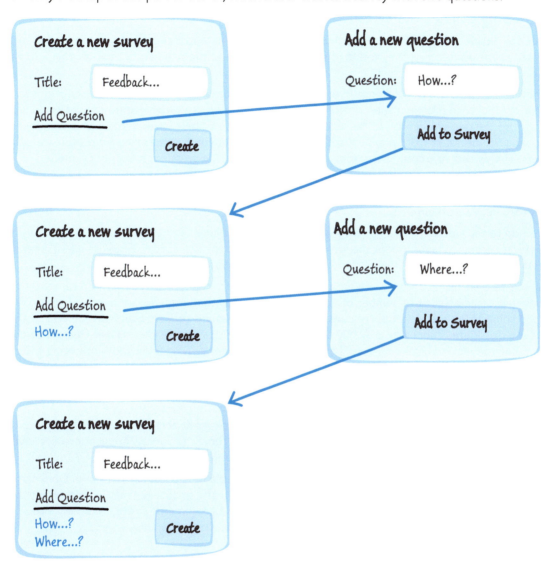

FIGURE 4
Creating a survey with two questions

As you can see in the diagram, this scenario involves two different screens that require the application to maintain state as the user adds questions to the survey.

The Surveys application must maintain session state while a user designs a survey.

OVERVIEW OF THE SOLUTION

This section describes the approach taken by Tailspin to meet the goals and requirements that relate to partitioning the application.

Partitioning Queues and Worker Roles

In order to enable giving priority to premium subscribers in the application Tailspin considered two alternatives for partitioning the workload for the standard and premium subscribers in the Tailspin Surveys worker role.

The first option is to use two different worker roles, one for tenants with standard subscriptions and one for tenants with premium subscriptions. Tailspin could then use two separate queues to deliver messages to the two worker roles. The second option is to use a single worker role with two queues, one queue for messages from premium subscribers and one queue for messages from standard subscribers. The worker role could then prioritize messages from the premium subscriber's queue.

Tailspin preferred the second option because it avoids the need to manage and run different types of worker role. It can modify the priorities of the different subscriptions by adjusting configuration values for a single worker role.

In addition to the worker role using two (or more) different queues to enable it to partition the work for different groups of subscribers, the web role must choose the correct queue to use when it sends a message to the worker role.

However, Tailspin realized that there are limitations in the throughput of Windows Azure storage queues that could affect the number of concurrent messages that a queue can handle. The recommended solution is to use multiple queues and implement a round-robin process at each end of the queues to distribute the messages evenly between the queues, and to read messages from all of the queues. See the section "Azure Queues Throughput" in Chapter 7, "Managing and Monitoring Multitenant Applications," for more information.

Tenant Isolation in Web Roles

Chapter 3, "Choosing a Multi-Tenant Data Architecture," describes how the Surveys application data model partitions the data by subscriber. This section describes how the Surveys application uses MVC routing tables and areas to make sure that a subscriber sees only his or her own data.

Tailspin considered using host headers and virtual sites to enable tenants to use their own DNS name to provide access to their public surveys. However, because many smaller tenants will not want the additional complexity of managing DNS entries, and because you cannot configure a new site and host header without redeploying the application, it decided against using this option.

The developers at Tailspin decided to use the path in the application's URL to indicate which subscriber is accessing the application. For the public Surveys website, the application doesn't require authentication.

The URL path identifies the functional area in the application, the subscriber, and the action.

The following are three sample paths on the Subscriber website:

- /survey/adatum/newsurvey
- /survey/adatum/newquestion
- /survey/adatum

The following are two example paths on the public Surveys website:

- /survey/adatum/launch-event-feedback
- /survey/adatum/launch-event-feedback/thankyou

The application uses the first element in the path to indicate the different areas of functionality within the application. In the initial release of the Tailspin Surveys service the only functional area is survey, but in the future Tailspin expects there to be additional functional areas such as onboarding and security. The second element indicates the subscriber name, in these examples "Adatum," and the last element indicates the action to perform, such as creating a new survey or adding a question to a survey.

You should take care when you design the path structure for your application that there is no possibility of name clashes that result from a value entered by a subscriber. In the Surveys application, if a subscriber creates a survey named "newsurvey" the path to this survey is the same as the path to the page subscribers use to create new surveys. However, the application hosts surveys on an HTTP endpoint and the page to create surveys on an HTTPS endpoint, so there is no name clash in this particular case.

> *A slug name is a string where all whitespace and invalid characters are replaced with a hyphen (-). The term comes from the newsprint industry and has nothing to do with those things in your garden!*

The third example element of the public Surveys website, "launch-event-feedback," is a slug name version of the survey title, originally "Launch Event Feedback," to make it URL friendly.

DNS Names, Certificates, and SSL in the Surveys Application

In Chapter 1, "The Tailspin Scenario," you saw how the Surveys application has three different groups of users. This section describes how Tailspin can use Domain Name System (DNS) entries to manage the URLs that each group can use to access the service, and how Tailspin plans to use SSL to protect some elements of the Surveys application.

To make it easy for the Surveys application to meet the requirements outlined earlier, the developers at Tailspin decided to use separate web roles. One web role will contain the subscriber and administrative functionality, while a separate web role will host the surveys themselves. This partitioning of the UI functionality enables Tailspin to scale each web role to support its usage profile independently of the other.

Having multiple web roles in a hosted cloud service affects the choice of URLs that you can use to access the application. Windows Azure assigns a single DNS name (for example, tailspin.cloudapp.net) to a cloud service, which means that different websites within a hosted service must have different port numbers. For example, two websites within Tailspin's hosted service could have the addresses listed in the following table.

Site A	Site B
http://tailspin.cloudapp.net:80	http://tailspin.cloudapp.net:81

> *You can use DNS **CNAME** records to map custom domain names to the default DNS names provided by Windows Azure. You can also use DNS A records to map a custom domain name to your service, but the IP address is only guaranteed to remain the same while the application is deployed. If you delete the deployment and then redeploy to the same cloud service, your application will have new IP address, and you will need to change the A record. An IP address is associated with a deployment, not a cloud service. For more information, see the blog post "Windows Azure Deployments and The Virtual IP."*

Because of the specific security requirements of the Surveys application, Tailspin decided to use the following URLs:

- https://tailspin.cloudapp.net
- http://tailspin.cloudapp.net

The next sections describe each of these.

https://tailspin.cloudapp.net

This HTTPS address uses the default port 443 to access the web role that hosts the administrative functionality for both subscribers and Tailspin. Because an SSL certificate protects this site, it is possible to map only a single custom DNS name. Tailspin plans to use an address such as *https://surveys.tailspin.com* to access this site.

To use HTTPS you must deploy a certificate to your cloud service. To avoid warnings in the browser, you should use an SSL certificate issued by a trusted third-party.

http://tailspin.cloudapp.net

This HTTP address uses the default port 80 to access the web role that hosts the public surveys. Because there is no SSL certificate, it is possible to map multiple DNS names to this site. Tailspin will configure a default DNS name such as *http://surveys.tailspin.com* to access the surveys, and individual tenants can then create their own CNAME entries to map to http://surveys.tailspin.com; for example, *http://surveys. adatum.com*, *http://surveys.tenant2.org*, or *http://survey.tenant3.co.de*.

Accessing Tailspin Surveys in Different Geographic Regions

Tailspin plans to create separate hosted cloud services to host copies of the Surveys service in different geographic regions. This will enable subscribers to choose where to host their surveys in order to minimize any latency for their users. Each regional version of the public Tailspin Surveys service will be available at a different URL by using a different subdomain. For example, Tailspin could use the following URLs to enable access to versions of Tailspin Surveys hosted in the US, Europe, and the Far East: *http://surveys.tailspin.com*, *http://eusurveys.tailspin. com*, and *http://fesurveys.tailspin.com*. Subscribers could then map their own DNS names to these addresses.

Tailspin will need to publish some guidance to subscribers that describes how they can set up their CNAMEs in their DNS settings.

> *If Tailspin wanted to enable a subscriber to publish a survey that it intends to be available globally, rather than in a specific region, the survey could be hosted on all the Tailspin Surveys cloud services. Tailspin could then use Windows Azure Traffic Manager to route client requests to closest version of Tailspin Surveys. For more information see "Traffic Manager" and Chapter 6, "Maximizing Scalability, Availability, and Performance in the Orders Application," of the related patterns & practices guide "Building Hybrid Applications in the Cloud on Windows Azure."*

Maintaining Session State

The tenant website uses session state during the survey creation process. The developers at Tailspin considered three options for managing session state.

- Use JavaScript and manage the complete survey creation workflow on the client. Then use AJAX calls to send the complete survey to the server after it's complete.

- Use the standard built-in **Request.Session** object to store the intermediate state of the survey while the user is creating it. Because the Tailspin web role will run on several node instances Tailspin cannot use the default, in-memory session state provider, and would have to use another provider such as the session state provider that's included in Windows Azure Caching. For more information, see *"Caching in Windows Azure."*

• Use an approach similar to ViewState that serializes and deserializes the workflow state and passes it between the two pages.

You can compare the three options using several different criteria. Which criteria are most significant will depend on the specific requirements of your application.

Simplicity

Something that is simple to implement is usually also easy to maintain. The first option is the most complex of the three, requiring JavaScript skills and good knowledge of an AJAX library. It is also difficult to unit test. The second option is the easiest to implement because it uses the standard ASP.NET Session object. Using the session state provider is simply a matter of "plugging-in" the Windows Azure Caching Service session state provider in the Web.config file. The third option is moderately complex, but you can simplify the implementation by using some of the features in ASP.NET MVC. Unlike the second option, it doesn't require any server side setup or configuration other than the standard MVC configuration.

If, in the future, Tailspin decides to use Windows Azure Caching for other purposes in the Surveys application, this could lead to greater pressure on the cache and increase the likelihood of cache evictions.

Data is encoded using a Base64 algorithm when you store it in the ASP. NET ViewState, so any estimate of the average question size must take this into account.

Although the second option is easy to implement, it does introduce some potential concerns about the long-term maintenance of the application. The current version of Windows Azure Caching does not support disabling eviction on a cache, so if the cache fills up it could evict session data while the session is still active. The cache uses a least recently used (LRU) policy if it needs to evict items. Tailspin should monitor cache usage and check for the situation where the cache has evicted items from an active session. If this occurs, Tailspin can increase the size of the cache or enable compression to store more data in the existing cache.

Cost

The first option has the lowest costs because it uses a single POST message to send the completed survey to the server. The second option has moderate costs. If Tailspin chooses to use a Windows Azure Shared Cache, it is easy to determine the cost because Windows Azure bills for the cache explicitly based on the cache size. If Tailspin chooses to use Windows Azure Caching, it is harder to quantify the cost because this type of cache uses a proportion of the memory in Tailspin's role instances. The third option incurs costs that arise from bandwidth usage; Tailspin can estimate the costs based on the expected number of questions created per day and the average size of the questions.

Performance

The first option offers the best performance because the client performs almost all the work with no roundtrips to the server until the browser sends a final HTTP POST message containing the complete survey. The second option will introduce a small amount of latency into the application; the amount of latency will depend on the number of concurrent sessions, the amount of data in the session objects, and the latency between the web role and the cache. If Tailspin uses Windows Azure Shared Caching, the latency between the web role and the cache maybe greater than if Tailspin uses Windows Azure Caching. The third option will also introduce some latency because each question will require a round-trip to the server and each HTTP request and response message will include all the current state data.

Scalability

All three options scale well. The first option scales well because it doesn't require any session state data outside the browser, the second and third options scale well because they are "web-farm friend-ly" solutions that you can deploy on multiple web roles.

Robustness

The first option is the least robust, relying on client-side JavaScript code. The second option is robust, using a feature that is a standard part of the Windows Azure. The third option is also robust, using easily testable server-side code.

User Experience

The first option provides the best user experience because there are no postbacks during the survey creation process. The other two options require a postback for each question.

Security

The first two options offer good security. With the first option, the browser holds all the survey in memory until the survey creation is complete, and with the second option, the browser just holds a cookie with a session ID while Windows Azure Caching holds the survey data. The third option is not so secure because it simply serializes the data to Base64 without encrypting it. It's possible that sensi-tive data could "leak" during the flow between pages.

Tailspin decided to use the second option that uses the session state provider included with Windows Azure Caching. This solution meets Tailspin's criteria for this part of the Tailspin Surveys application.

Isolating Cached Tenant Data

In addition to using a Windows Azure cache for storing session state, Tailspin also chose to use Windows Azure Caching to cache application data. Tailspin chose to use a co-located Windows Azure cache that uses a proportion of the memory available to the public web site web role instances.

> *For more information about Windows Azure Caching, see "Overview of Caching in Windows Azure."*

In order to isolate tenant data in the cache that Tailspin Surveys uses to cache frequently used data, such as survey definitions and tenant configuration data, the developers at Tailspin chose to use regions in the Windows Azure cache and assign each tenant its own region. To retrieve an item from the cache the calling code must specify the cache region that contains the required item. This makes it easy for the application to ensure that only data that belongs to a tenant is accessed by that tenant.

INSIDE THE IMPLEMENTATION

Now is a good time to walk through some of the code in the Tailspin Surveys application in more detail. As you go through this section, you may want to download the Visual Studio solution for the Tailspin Surveys application from *http://wag.codeplex.com/*.

Prioritizing Work in a Worker Role

To enable the worker role in Tailspin Surveys to support prioritizing work from different groups of tenants, the developers at Tailspin introduced some "plumbing" code to launch tasks within a worker role. The following code sample from the **Run** method of the **WorkerRole** class in the Tailspin.Workers. Surveys project shows how the Surveys application uses the **BatchMultipleQueueHandler** class from this plumbing code.

```C#
var standardQueue = this.container.Resolve
        <IAzureQueue<SurveyAnswerStoredMessage>>
          (SubscriptionKind.Standard.ToString());
var premiumQueue = this.container.Resolve
        <IAzureQueue<SurveyAnswerStoredMessage>>
          (SubscriptionKind.Premium.ToString());


BatchMultipleQueueHandler
  .For(premiumQueue, GetPremiumQueueBatchSize())
  .AndFor(standardQueue, GetStandardQueueBatchSize())
  .Every(TimeSpan.FromSeconds(
                    GetSummaryUpdatePollingInterval()))
  . WithLessThanTheseBatcheIterationsPerCycle(
                    GetMaxBatchIterationsPerCycle())
  .Do(this.container.Resolve
      <UpdatingSurveyResultsSummaryCommand>());
```

> The **For**, **AndFor**, **Every**, **WithLessThanTheseBatchesPerCycle**, and **Do** *methods implement a fluent API for instantiating tasks in the worker role. Fluent APIs help to make the code more legible.*

The **Run** method creates two Windows Azure queues, one to handle messages for standard subscribers and one to handle messages for premium subscribers. The worker role prioritizes processing messages for premium subscribers based on the batch sizes it reads from the service configuration file using the **GetPremiumQueueBatchSize** and **GetStandardQueueBatchSize** methods. The worker role also uses the **GetSummaryUpdatePollingInterval** method to read the service configuration file and set the polling interval for reading messages from the queue, and the **GetMaxBatchIterationsPerCycle** method to set the maximum number of messages that will be processed in each cycle.

> It's important to limit the maximum number of messages that the worker role process can read from the queue in each cycle. If the code continues reading messages until the queue is empty, but the web role is adding messages faster than the web role can process them, the cycle will never end!

The **BatchMultipleQueueHandler** class in the worker role enables you to invoke commands of type **IBatchCommand<T>** by using the **Do** method. You can invoke these commands on several Windows Azure queues of type **IAzureQueue** by using the **For** and **AndFor** methods, at an interval specified by the **Every** method. The **WithLessThanTheseBatchIterationsPerCycle** method limits the number of batches that the task retrieves from the queue before it processes the messages.

The example code you saw above shows the worker role processing a premium and a standard queue, both of which transport **SurveyAnswerStoredMessage** messages. It processes the messages every ten seconds by using the **UpdatingSurveyResultsSummaryCommand** class.

There is also a **QueueHandler** class that processes messages from a single queue. It has a slightly simpler API; the **Do** method enables you to invoke commands of type **ICommand**, the **For** method identifies a single Windows Azure queue of type **IAzureQueue**, and the **Every** method specifies how frequently to process messages.

> *The tasks that Tailspin runs in the worker role using the task framework described in this chapter include saving the survey responses and calculating the summary statistics. See Chapter 5, "Maximizing Availability, Scalability, and Elasticity," for descriptions of these tasks.*

The design of the **BatchMultipleQueueHandler** class enables the compiler to check that the two queues and the **UpdatingSurveyResultsSummaryCommand** class all use the same message type.

The BatchMultipleQueueHandler and the Related Classes

This section describes the implementation of the **BatchMultipleQueueHandler** class and the related classes. The implementation using the **QueueHandler** class is very similar but runs tasks that implement the simpler **ICommand** interface instead of the **IBatchCommand** interface.

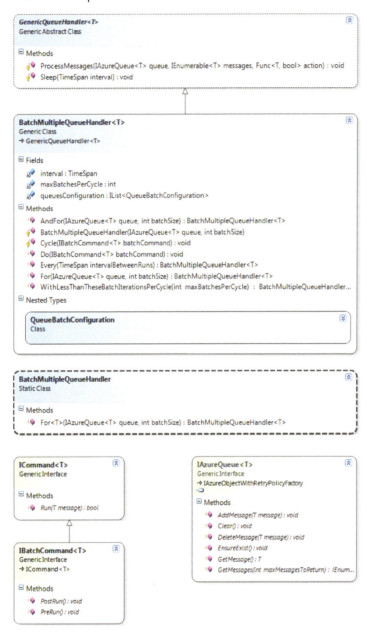

FIGURE 5
Key plumbing types

Figure 5 shows the key types that make up the plumbing code related to the **BatchMultipleQueue-Handler** class that the application uses to prioritize work for premium subscribers. The worker role first invokes the **For** method in the static **BatchMultipleQueueHandler** class, which invokes the **For** method in the **BatchMultipleQueueHandler<T>** class. The **For** method returns a **BatchMultiple-QueueHandler<T>** instance that contains a reference to the **IAzureQueue<T>** instance to monitor.

The plumbing code identifies the queue by name and associates it with a message type that derives from the **AzureQueueMessage** type. For example, both the Standard and Premium queues handle **SurveyAnswerStoredMessage** messages. The following code example shows how the static **For** method in the **BatchMultipleQueueHandler** class instantiates a **BatchMultipleQueueHandler<T>** instance and invokes the **For** method, passing the required batch size as a parameter.

```C#
using Tailspin.Web.Survey.Shared.Stores.AzureStorage;

public static class BatchMultipleQueueHandler
{
  public static BatchMultipleQueueHandler<T>
    For<T>(IAzureQueue<T> queue, int batchSize)
    where T : AzureQueueMessage
  {
    return BatchMultipleQueueHandler<T>.For
      (queue, batchSize);
  }
}
```

Next, the worker role invokes the **AndFor** method for each additional queue that transports messages. The following code sample shows both the **For** and the **AndFor** methods of the **BatchMultiple-QueueHandler<T>** class.

```C#
public static BatchMultipleQueueHandler<T> For
              (IAzureQueue<T> queue, int batchSize)
{
  if (queue == null)
  {
    throw new ArgumentNullException("queue");
  }

  batchSize = Math.Max(1, batchSize);
  return new BatchMultipleQueueHandler<T>(queue, batchSize);
}
```

```csharp
public BatchMultipleQueueHandler<T> AndFor
                (IAzureQueue<T> queue, int batchSize)
{
  if (queue == null)
  {
    throw new ArgumentNullException("queue");
  }

  batchSize = Math.Max(1, batchSize);
  this.queuesConfiguration.Add
    (QueueBatchConfiguration.BuildConfig(queue, batchSize));
  return this;
}
```

Next, the worker role invokes the **Every** method of the **Batch-MultipleQueueHandler<T>** object to specify how frequently the task should be run. Then it invokes the **WithLessThanTheseBatch-IterationsPerCycle** method to limit the number of batches to process in each cycle.

Use Task.Factory.StartNew in preference to ThreadPool. QueueUserWorkItem to ensure that your application can maximize performance on any system on which it will run.

Finally, the worker role invokes the **Do** method of the **BatchMultiple-QueueHandler<T>** object, passing an **IBatchCommand** object that identifies the command that the plumbing code should execute on each message in the queue. The following code example shows how the **Do** method uses the **Task.Factory.StartNew** method from the Task Parallel Library (TPL) to execute the **PreRun**, **ProcessMessages**, and **PostRun** methods on the queue at the requested interval.

```csharp
C#
public virtual void Do(IBatchCommand<T> batchCommand)
{
  Task.Factory.StartNew(
  () =>
  {
    while (true)
    {
      this.Cycle(batchCommand);
    }
  },
  TaskCreationOptions.LongRunning);
}
```

```
protected void Cycle(IBatchCommand<T> batchCommand)
{
  try
  {
    batchCommand.PreRun();

    int batches = 0;
    bool continueProcessing;
    do
    {
      continueProcessing = false;
      foreach (var queueConfig in this.queuesConfiguration)
      {
        var messages = queueConfig.Queue
          .GetMessages(queueConfig.BatchSize);
        GenericQueueHandler<T>.ProcessMessages(
          queueConfig.Queue, messages, batchCommand.Run);
        continueProcessing |= messages.Count()
          >= queueConfig.BatchSize;
      }
      batches++;
    }
    while (continueProcessing && batches
            < maxBatchesPerCycle);

    batchCommand.PostRun();

    this.Sleep(this.interval);
  }
  catch (TimeoutException ex)
  {
    TraceHelper.TraceWarning(ex.TraceInformation());
  }
  catch (Exception ex)
  {
    // No exception should get here -
    // we don't want the handler to stop
    // (we log it as ERROR)
    TraceHelper.TraceError(ex.TraceInformation());
  }
}
```

By configuring one queue to use a larger batch size, you can ensure that the worker role processes messages in that queue faster than other queues. In addition, reading messages from queues in batches can reduce your costs because it reduces the number of storage operations.

The **Cycle** method repeatedly pulls messages for processing from the queue, up to the number specified as the batch size in the **For** and **AndFor** methods, in a single transaction; until there are no more messages left or the maximum batches per cycle is reached.

The following code example shows the **ProcessMessages** method in the **GenericQueueHandler** class that performs the actual message processing.

```csharp
C#
protected static void ProcessMessages(IAzureQueue<T> queue,
        IEnumerable<T> messages, Func<T, bool> action)
{
  ...

  foreach (var message in messages)
  {
    var allowDelete = false;
    var corruptMessage = false;

    try
    {
      allowDelete = action(message);
    }
    catch (Exception ex)
    {
      TraceHelper.TraceError(ex.TraceInformation());
      allowDelete = false;
      corruptMessage = true;
    }
    finally
    {
      if (allowDelete || (corruptMessage
        && message.GetMessageReference().DequeueCount > 5))
      {
        queue.DeleteMessage(message);
      }
    }
  }
}
```

This method uses the action parameter to invoke the custom command on each message in the queue; if this fails it logs the error. Finally, the method checks for poison messages by looking at the **Dequeue-Count** property of the message; if the application has tried more than five times to process the message, the method deletes the message.

> *Instead of deleting poison messages, you should send them to a dead message queue for analysis and troubleshooting.*

Using MVC Routing Tables

The request routing implementation in the Tailspin Surveys application uses a combination of ASP.NET routing tables and MVC areas to identify the subscriber and map requests to the correct functionality within the application.

The following code example shows how the public Surveys Web site uses routing tables to determine which survey to display based on the URL.

```
C#
using System.Web.Mvc;
using System.Web.Routing;

public static class AppRoutes
{
  public static void RegisterRoutes(RouteCollection routes)
  {
    routes.MapRoute(
      "Home",
      string.Empty,
      new { controller = "Surveys", action = "Index" });

    routes.MapRoute(
      "ViewSurvey",
      "survey/{tenant}/{surveySlug}",
      new { controller = "Surveys", action = "Display" });

    routes.MapRoute(
      "ThankYouForFillingTheSurvey",
      "survey/{tenant}/{surveySlug}/thankyou",
      new { controller = "Surveys", action = "ThankYou" });
  }
}
```

The code extracts the tenant name and survey name from the URL and passes them to the appropriate action method in the **Surveys-Controller** class. The following code example shows the **Display** action method that handles HTTP **GET** requests.

```
C#
[HttpGet]
public ActionResult Display(string tenant,
                            string surveySlug)
{
  var surveyAnswer = CallGetSurveyAndCreateSurveyAnswer(
          this.surveyStore, tenant, surveySlug);

  var model =
      new TenantPageViewData<SurveyAnswer>(surveyAnswer);
  if (surveyAnswer != null)
  {
    model.Title = surveyAnswer.Title;
  }
  return this.View(model);
}
```

If the user requests a survey using a URL with a path value of /survey/adatum/launch-event-feedback, the value of the *tenant* parameter will be "Adatum" and the value of the *surveySlug* parameter will be "launch-event-feedback." The **Display** action method uses the parameter values to retrieve the survey definition from the store, populate the model with this data, and pass the model to the view that renders it to the browser.

There is also a Display action to handle HTTP POST requests. This controller action is responsible for saving the data from a filled out survey.

The Subscriber website is more complex because, in addition to enabling subscribers to design new surveys and analyze survey results, it must handle authentication and onboarding new subscribers. Because of this complexity it uses MVC areas as well as a routing table. The following code from the **AppRoutes** class in the TailSpin.Web project shows how the application maps top level requests to the controller classes that handle onboarding and authentication.

```C#
public static void RegisterRoutes(RouteCollection routes)
{
  routes.MapRoute(
    "OnBoarding",
    string.Empty,
    new { controller = "OnBoarding", action = "Index" });

  routes.MapRoute(
    "FederationResultProcessing",
    "FederationResult",
    new { controller = "ClaimsAuthentication",
              action = "FederationResult" });

  routes.MapRoute(
    "FederatedSignout",
    "Signout",
    new { controller = "ClaimsAuthentication",
              action = "Signout" });
  }
  ...
}
```

MVC areas enable you to group multiple controllers together within the application, making it easier to work with large MVC projects. Each MVC area typically represents a different functional area within the application.

The application also defines an MVC area for the core survey functionality. MVC applications register areas by calling the **RegisterAllAreas** method. In the TailSpin.Web project you can find this call in the **Application_Start** method in the Global.asax.cs file. The **RegisterAllAreas** method searches the application for classes that extend the **AreaRegistration** class, and then it invokes the **RegisterArea** method. The following code example shows a part of this method in the **SurveyAreaRegistration** class.

```csharp
C#
public override void RegisterArea(
                      AreaRegistrationContext context)
{
  context.MapRoute(
    "MySurveys",
    "survey/{tenant}",
    new { controller = "Surveys", action = "Index" });

  context.MapRoute(
    "NewSurvey",
    "survey/{tenant}/newsurvey",
    new { controller = "Surveys", action = "New" });

  context.MapRoute(
    "NewQuestion",
    "survey/{tenant}/newquestion",
    new { controller = "Surveys", action = "NewQuestion" });

  context.MapRoute(
    "AddQuestion",
    "survey/{tenant}/newquestion/add",
    new { controller = "Surveys", action = "AddQuestion" });

  ...
}
```

Notice how all the routes in this routing table include the tenant name that MVC passes as a parameter to the controller action methods.

Web Roles in Tailspin Surveys

To implement the two different websites within a single hosted cloud service, the developers at Tailspin defined two web roles in the solution. The first website, named TailSpin.Web, is an MVC project that handles the administrative functionality within the application. This website requires authentication and authorization, and users access it using HTTPS. The second website, named Tailspin.Web.Survey.Public, is an MVC project that handles users filling out surveys. This website is public, and users access it using HTTP.

The following code example shows the contents of an example ServiceDefinition.csdef file and the definitions of the two web roles in Tailspin Surveys:

```XML
<?xml version="1.0" encoding="utf-8"?>
<ServiceDefinition name="Tailspin.Cloud" xmlns=...>
  <WebRole name="Tailspin.Web"
    enableNativeCodeExecution="true">
    <Sites>
      <Site name="Web">
        <Bindings>
          <Binding name="HttpsIn" endpointName="HttpsIn" />
        </Bindings>
      </Site>
    </Sites>
    <Certificates>
      <Certificate name="localhost"
        storeLocation="LocalMachine" storeName="My" />
    </Certificates>
    <Endpoints>
      <InputEndpoint name="HttpsIn" protocol="https"
        port="443" certificate="localhost" />
    </Endpoints>
    ...
  </WebRole>
  <WebRole name="Tailspin.Web.Survey.Public">
    <Sites>
      <Site name="Web">
        <Bindings>
          <Binding name="HttpIn" endpointName="HttpIn" />
        </Bindings>
      </Site>
    </Sites>
    <Endpoints>
      <InputEndpoint name="HttpIn" protocol="http"
        port="80" />
    </Endpoints>
    ...
  </WebRole>
  <WorkerRole name="Tailspin.Workers.Surveys">
    ...
  </WorkerRole>
</ServiceDefinition>
```

This example ServiceDefinition.csdef file does not exactly match the file in the downloadable solution, which uses a different name for the SSL certificate.

> *Remember, you may want to use different SSL certificates when you are testing the application using the local compute emulator. You must make sure that the configuration files reference the correct certificates before you publish the application to Windows Azure. For more information about managing the deployment, see Chapter 3, "Moving to Windows Azure Cloud Services," in the guide "Moving Applications to the Cloud."*

In addition to the two web role projects, the solution also contains a worker role project and a library project named TailSpin.Web.Survey.Shared that contains code shared by the web and worker roles. This shared code includes the model classes and the data access layer.

Implementing Session Management

The Surveys application must maintain some state data for each user as they design a survey. This section describes the design and implementation of user state management in the Surveys application.

The following code examples shows how the action methods in the **SurveysController** controller class in the TailSpin.Web project use the MVC **TempData** property to cache the survey definition while the user is designing a new survey. Behind the scenes, the **TempData** property uses the ASP.NET session object to store cached objects.

The **New** method that handles **GET** requests, shown here, is invoked when a user navigates to the **New Survey** page.

```csharp
C#
[HttpGet]
public ActionResult New()
{
  var cachedSurvey = (Survey)this.TempData[CachedSurvey];

  if (cachedSurvey == null)
  {
    cachedSurvey = new Survey();  // First time to the page
  }

  var model = this.CreateTenantPageViewData(cachedSurvey);
  model.Title = "New Survey";

  this.TempData[CachedSurvey] = cachedSurvey;

  return this.View(model);
}
```

PARTITIONING MULTI-TENANT APPLICATIONS

The **NewQuestion** method is invoked when a user chooses the **Add Question** link on the **Create a new survey** page. The method retrieves the cached survey that the **New** method created, ready to display it to the user.

```csharp
C#
[HttpPost]
[ValidateAntiForgeryToken]
public ActionResult NewQuestion(Survey contentModel)
{
  var cachedSurvey = (Survey)this.TempData[CachedSurvey];

  if (cachedSurvey == null)
  {
    return this.RedirectToAction("New");
  }

  cachedSurvey.Title = contentModel.Title;
  this.TempData[CachedSurvey] = cachedSurvey;

  var model = this.CreateTenantPageViewData(new Question());
  model.Title = "New Question";

  return this.View(model);
}
```

The **AddQuestion** method is invoked when a user chooses the **Add to survey** button on the **Add a new question** page. The method retrieves the cached survey and adds the new question, then updates the survey stored in the session.

```csharp
C#
[HttpPost]
[ValidateAntiForgeryToken]
public ActionResult AddQuestion(Question contentModel)
{
  var cachedSurvey = (Survey)this.TempData[CachedSurvey];

  if (!this.ModelState.IsValid)
  {
    this.TempData[CachedSurvey] = cachedSurvey;
    var model = this.CreateTenantPageViewData(
                contentModel ?? new Question());
    model.Title = "New Question";
    return this.View("NewQuestion", model);
  }

  if (contentModel.PossibleAnswers != null)
  {
    contentModel.PossibleAnswers =
      contentModel.PossibleAnswers.Replace("\r\n", "\n");
  }

  cachedSurvey.Questions.Add(contentModel);
  this.TempData[CachedSurvey] = cachedSurvey;
  return this.RedirectToAction("New");
}
```

The **New** method that handles **POST** requests is invoked when a user chooses the **Create** button on the **Create a new survey** page. The method retrieves the completed, cached survey, saves it to persistent storage, and removes it from the session.

```C#
[HttpPost]
[ValidateAntiForgeryToken]
public ActionResult New(Survey contentModel)
{
  var cachedSurvey = (Survey)this.TempData[CachedSurvey];

  if (cachedSurvey == null)
  {
    return this.RedirectToAction("New");
  }

  if (cachedSurvey.Questions == null ||
      cachedSurvey.Questions.Count <= 0)
  {
    this.ModelState.AddModelError("ContentModel.Questions",
      string.Format(CultureInfo.InvariantCulture,
      "Please add at least one question to the survey."));
  }

  contentModel.Questions = cachedSurvey.Questions;
  if (!this.ModelState.IsValid)
  {
    var model = this.CreateTenantPageViewData(contentModel);
    model.Title = "New Survey";
    this.TempData[CachedSurvey] = cachedSurvey;
    return this.View(model);
  }

  contentModel.Tenant = this.TenantName;
  try
  {
    this.surveyStore.SaveSurvey(contentModel);
  }
  catch (DataServiceRequestException ex)
  {
    ...
  }

  this.TempData.Remove(CachedSurvey);
  return this.RedirectToAction("Index");
}
```

Tailspin use the TempData property instead of working with the ASP.NET Session object directly because the entries in the TempData dictionary live only for a single request, after which they're automatically removed from the session. This makes it easier to manage the contents of the session.

Tailspin just needed to modify the configuration settings in the Surveys application to change from using the default, in-memory, ASP.NET session state provider to using the session state provider that uses Windows Azure Caching. No application code changes were necessary. The following sections describe the configuration changes Tailspin made.

Configuring a Cache in Windows Azure Caching

Tailspin uses the ASP.NET 4 Caching Session State Provider in the tenant web role. This requires Tailspin to configure Windows Azure Caching in the project, and Tailspin chose to use a co-located cache that uses a proportion of the web role's memory. You can configure the settings for this type of cache using the role properties in Visual Studio.

The following sample shows the part of the service configuration file where the cache configuration is stored. The value for **NamedCaches** is the default set by the SDK; it allows you to change the cache settings while the application is running simply by editing the configuration file.

```XML
<ServiceConfiguration serviceName="Tailspin.Cloud" ... >
  <Role name="Tailspin.Web">
    <Instances count="1" />
    <ConfigurationSettings>
      ...
      <Setting
        name="Microsoft.WindowsAzure.Plugins
              .Caching.NamedCaches"
        value="{&quot;caches&quot;:[{&quot;name&quot;
              :&quot;default&quot;,&quot;policy&quot;
              :{&quot;eviction&quot;
              :{&quot;type&quot;:0},&quot;expiration&quot;
              :{&quot;defaultTTL&quot;
              :10,&quot;isExpirable&quot;
              :true,&quot;type&quot;:1},&quot;
               serverNotification&quot;
              :{&quot;isEnabled&quot;
              :false}},&quot;secondaries&quot;:0}]}" />
      <Setting
        name="Microsoft.WindowsAzure.Plugins
              .Caching.DiagnosticLevel"
        value="1" />
      <Setting name="Microsoft.WindowsAzure.Plugins
                    .Caching.Loglevel"
              value="" />
      <Setting name="Microsoft.WindowsAzure.Plugins
                    .Caching.CacheSizePercentage"
              value="30" />
```

```xml
      <Setting name="Microsoft.WindowsAzure.Plugins
                     .Caching.ConfigStoreConnectionString"
             value="UseDevelopmentStorage=true" />
   </ConfigurationSettings>
   ...
  </Role>
  <Role name="Tailspin.Web.Survey.Public">
    ...
  </Role>
  <Role name="Tailspin.Workers.Surveys">
    ...
  </Role>
</ServiceConfiguration>
```

This example shows how to configure a default cache that use 30% of the available memory in the Tailspin.Web web role instances. It uses the local storage emulator to store the cache's runtime state, and you must change this to use a Windows Azure storage account when you deploy the application to Windows Azure.

> *Tailspin used NuGet to add the required assemblies and references to the Tailspin.Web.Survey. Shared project.*

Configuring the Session State Provider in the TailSpin.Web Application

The final changes that Tailspin made were to the Web.config file in the TailSpin.Web project. The following example shows these changes.

```xml
XML
<configSections>
  ...
  <section name="dataCacheClients"
    type="Microsoft.ApplicationServer
          .Caching.DataCacheClientsSection,
          Microsoft.ApplicationServer.Caching.Core"
    allowLocation="true" allowDefinition="Everywhere"/>
</configSections>
...
<dataCacheClients>
    <tracing sinkType="DiagnosticSink"
             traceLevel="Error" />
    <dataCacheClient name="default"
                     maxConnectionsToServer="5">
      <autoDiscover isEnabled="true"
                    identifier="Tailspin.Web" />
    </dataCacheClient>
  </dataCacheClients>
  ...
```

```
<system.web>
  <sessionState mode="Custom"
      customProvider="AppFabricCacheSessionStoreProvider">
    <providers>
      <add name="AppFabricCacheSessionStoreProvider"
        type="Microsoft.Web.DistributedCache
              .DistributedCacheSessionStateStoreProvider,
              Microsoft.Web.DistributedCache"
        cacheName="default" useBlobMode="false"
        dataCacheClientName="default" />
    </providers>
  </sessionState>
  ...
</system.web>
```

The **sessionState** section configures the application to use the default cache provided by the Windows Azure Caching session state provider.

Caching Frequently Used Data

The public website frequently accesses survey definitions and tenant data in read-only mode to display surveys to respondents. To reduce latency the application attempts to use cached versions of this data if it is available.

The following code sample shows the **TenantCacheHelper** class that ensures that each tenant has its own region in the cache in order to isolate its data from other tenants. The sample also shows how the **RemoveAllFromCache** method removes all the cache entries that belong to a single tenant.

```csharp
C#
internal static class TenantCacheHelper
{
  private static readonly DataCacheFactory CacheFactory;
  private static readonly IRetryPolicyFactory
                            RetryPolicyFactory;

  ...

  internal static void AddToCache<T>(string tenant,
                string key, T @object) where T : class
  {
    GetRetryPolicy().ExecuteAction(() =>
    {
      DataCache cache = CacheFactory.GetDefaultCache();
      if (!cache.GetSystemRegions().Contains(
                  tenant.ToLowerInvariant()))
```

```
    {
      cache.CreateRegion(tenant.ToLowerInvariant());
    }
    cache.Put(key.ToLowerInvariant(), @object,
                  tenant.ToLowerInvariant());
  });
}

internal static T GetFromCache<T>(string tenant,
          string key, Func<T> @default) where T : class
{
  return GetRetryPolicy().ExecuteAction<T>(() =>
  {
    var result = default(T);

    var success = false;
    DataCache cache = CacheFactory.GetDefaultCache();
    result = cache.Get(key.ToLowerInvariant(),
                      tenant.ToLowerInvariant()) as T;
    if (result != null)
    {
      success = true;
    }
    else if (@default != null)
    {
      result = @default();
      if (result != null)
      {
        AddToCache(tenant.ToLowerInvariant(),
                key.ToLowerInvariant(), result);
      }
    }
    TraceHelper.TraceInformation(
      "cache {2} for {0} [{1}]",
      key, tenant, success ? "hit" : "miss");
    return result;
  });
}

internal static void RemoveFromCache(string tenant,
                                    string key)
```

```
  {
    GetRetryPolicy().ExecuteAction(() =>
    {
      DataCache cache = CacheFactory.GetDefaultCache();
      cache.Remove(key.ToLowerInvariant(),
                   tenant.ToLowerInvariant());
    });
  }

  internal static void RemoveAllFromCache(string tenant)
  {
    GetRetryPolicy().ExecuteAction(() =>
    {
      DataCache cache = CacheFactory.GetDefaultCache();
      cache.RemoveRegion(tenant.ToLowerInvariant());
    });
  }

  ...
}
```

The following code sample shows how the **SurveyStore** class uses the **TenantCacheHelper** class to maintain survey definitions in the cache.

```
C#
public class SurveyStore : ISurveyStore
{
  ...
  public void SaveSurvey(Survey survey)
  {
    ...
    TenantCacheHelper.AddToCache(survey.Tenant,
                                 slugName, survey);
    ...
  }

  public void DeleteSurveyByTenantAndSlugName(
              string tenant, string slugName)
  {
    ...
    TenantCacheHelper.RemoveFromCache(tenant, slugName);
    ...
  }
```

```
public Survey GetSurveyByTenantAndSlugName(string tenant,
               string slugName, bool getQuestions)
{
  ...
  return this.CacheEnabled ?
    TenantCacheHelper.GetFromCache(tenant,
      slugName, resolver) : resolver();
  ...
}
...
}
```

MORE INFORMATION

All links in this book are accessible from the book's online bibliography available at: http://msdn.microsoft.com/library/jj871057.aspx.

For more information about Windows Azure multi-tenant application design, see *"Designing Multitenant Applications on Windows Azure."*

For more information about routing in ASP.NET, see *"ASP.NET Routing"* on MSDN.

For more information about using CNAME entries in DNS, see the post *"Custom Domain Names in Windows Azure"* on Steve Marx's blog.

For a description of the different caching options available in Windows Azure, see *"Caching in Windows Azure."*

For more information about Windows Azure resource provisioning, see *"Provisioning Windows Azure for Web Applications."*

For more information about the hard and soft limits in Windows Azure, see *"Best Practices for the Design of Large-Scale Services on Windows Azure Cloud Services"* on MSDN.

For more information about fluent APIs, see the entry for *"Fluent interface"* on Wikipedia.

For information about the Task Parallel Library, see *"Task Parallel Library"* on MSDN.

For information about the advantages of using the Task Parallel library instead of working with the thread pool directly, see the following:

- The article *"Optimize Managed Code for Multi-Core Machines"* in MSDN Magazine.
- The blog post *"Choosing Between the Task Parallel Library and the ThreadPool"* on the Parallel Programming with .NET blog.

5

Maximizing Availability, Scalability, and Elasticity

This chapter explores how you can maximize performance and availability for multi-tenant applications that run in Windows Azure. This includes considering how you can ensure that the application is scalable and responsive, and how you can take advantage of the elasticity available in Windows Azure to minimize running costs while meeting performance requirements.

Topics you will see discussed in this chapter include maximizing availability through geo-location, caching, and by using the Content Delivery Network (CDN); maximizing scalability through the use of Windows Azure storage queues, background processing, and asynchronous code; and implementing elasticity by controlling the number of web and worker role instances that are deployed and executing.

MAXIMIZING AVAILABILITY IN MULTI-TENANT APPLICATIONS

Multiple tenants sharing an instance of a role or other resource in Windows Azure increases the risk that the application becomes unavailable for several tenants. For example, if a multi-tenant worker role becomes unavailable it affects all of the tenants sharing the role, whereas the failure of a single-tenant worker role only affects that one tenant. These risks can increase if tenants have the ability to apply extensive customizations to the application. You must ensure that any extensions to the application, added by either the provider or a tenant, will not introduce errors that could affect the availability of the role.

In general, Windows Azure enables you to mitigate these risks by using multiple instances of resources. For example, you can run multiple instances of any web or worker role. Windows Azure will detect any failed role instances and route requests to other, functioning instances. Windows Azure will also attempt to restart failed instances.

One of the major advantages Windows Azure offers is the ability to use and pay for only what you actually need, while being able to increase and decrease the resources you use on demand without being forced to invest in spare or standby capacity.

113

However, running multiple instances of a role does impose some restrictions on your design. For example, because the Windows Azure load balancer can forward requests to any instance of a web role, either the role must be stateless or you must have a mechanism to share state across instances.

In the case of Windows Azure storage accounts, which contain blobs, tables, and queues, Windows Azure maintains multiple redundant copies. By default, Windows Azure uses geo-replication to make a copy of your data in another data center in addition to the multiple copies held in the data center that hosts your storage account.

> *For more information about how Windows Azure protects your data, see the blog post Introducing Geo-replication for Windows Azure Storage.*

It's important, whether your application is multi-tenant or single-tenant, that you understand the impact of a failure of any element of your application. In particular, you must understand which failure conditions Windows Azure can handle automatically, and which failure conditions your application or your administrators must handle.

Hosting copies of your Windows Azure application in multiple data-centers is another scenario that can help to keep your application available. For example, you can use Windows Azure Traffic Manager to define failover policies in the event that a deployment of your application in a particular datacenter becomes unavailable for some reason. However, you must still carefully plan how you will store your data and determine the data center or centers where you will store each item of data.

> *At the time of writing, Windows Azure Traffic Manager is a Community Technology Preview (CTP) release.*

MAXIMIZING SCALABILITY IN MULTI-TENANT APPLICATIONS

One of the reasons for running applications in Windows Azure is the scalability it offers. You can add resources to, or remove resources from your application as and when they are required. As mentioned previously, Windows Azure applications typically comprise multiple elements such as web and worker roles, storage, queues, virtual networks, and caches. One of the advantages of dividing the application into multiple elements is that you can then scale each element individually. You might be able to meet a change in your tenants' processing requirements by doubling the number of worker role instances without increasing the number of message queues, or the size of your cache.

You should also consider the granularity of the scalable elements of your Windows Azure application. For example, if you start with a small instance of a worker role rather than a large instance, you will have much finer control over the quantity of resources your application uses (and finer control over costs) because you can add or remove resources in smaller increments.

In a multi-tenant application, you may decide to allocate groups of tenants to specific resources. For example, you could have one worker role that is dedicated to handling premium tenants, and another worker role that is dedicated to handling standard tenants. In this way, you could scale the worker role that supports premium tenants independently. This might be useful if you have a different SLA for premium tenants to that for standard tenants.

In addition to running multiple instances of a role, Windows Azure offers some features such as caching that are specifically designed to enhance the scalability of your application.

Caching

One of the most significant things you can do to enhance the scalability of your Windows Azure application is to use caching. Typically, you should try to cache frequently accessed data from blob storage, table storage, and databases such as SQL Database. Caching can reduce the latency in retrieving data, reduce the workload on your storage system, and reduce the number of storage transactions.

However, you must consider issues such as how much caching space you will need, your cache expiration policies, your strategies for ensuring that the cache is loaded with the correct data, and how much staleness you are willing to accept. Appendix E, *"Maximizing Scalability, Availability, and Performance,"* in the guide *"Building Hybrid Applications in the Cloud on Windows Azure"* explores caching strategies for a range of scenarios.

If you decide to use a co-located Windows Azure cache in one or more of your role instances, you must consider the impact of allocating this memory to the cache instead of to the application.

In a multi-tenant application, you also need to consider how to isolate tenant data within the cache. For more information about how to partition a cache in Windows Azure, see Chapter 4, "Partitioning Multi-Tenant Applications."

Windows Azure offers two main caching mechanisms for application data: Windows Azure Shared Caching and Windows Azure Caching. For more information about the differences and similarities between these two approaches, see *"Overview of Caching in Windows Azure"* on MSDN.

SQL Database Federation

You can use SQL Database Federation to scale out your SQL Database databases across multiple servers. SQL Database federations work by horizontally partitioning the data stored in your SQL Database tables across multiple databases. For more information, see *"Federations in Windows Azure SQL Database."*

The type of horizontal partitioning used in SQL Database federations is often referred to as "sharding."

Shared Access Signatures

Shared Access Signatures (SAS) can help to make your application scalable by enabling clients to access items stored in blobs, tables, or queues directly and offloading the work of mediating access to these resources from your web and worker roles. For example, your application could use SAS to make the contents of a blob, such as an image, directly accessible from a web browser, without the need either to make the blob public, or the need to read private blob data in a web role and then pass it on to the client browser.

You could also use SAS to give a worker role hosted in another Windows Azure subscription access to specific rows in a table without either revealing your storage account keys or using a worker role in your subscription to retrieve the data on behalf of the worker role in the other subscription.

For more information about SAS, see *"Creating a Shared Access Signature"* on MSDN.

Content Delivery Network

The Content Delivery Network (CDN) can host static application resources such as media elements in edge caches. This reduces the latency for clients requesting these items, and it enhances the scalability of your application by offloading some of the work typically performed by web roles.

For more information about the CDN, see *"Caching"* on the Windows Azure features page.

IMPLEMENTING ELASTICITY IN MULTI-TENANT APPLICATIONS

Elasticity refers to the ability of the application to dynamically scale out or in based on actual or anticipated demand for resources. The discussion in the previous section about the scalability of web and worker roles in multi-tenant applications also applies to elasticity. In particular, you must decide at what level you want to enable elasticity: for individual tenants, for groups of tenants, or for all of the tenants in the application. You also need to identify which elements of the application, such as roles, storage, queues, and caches, must be elastic.

Elasticity is particularly important for multi-tenant applications because levels of demand may be less predictable than for single-tenant applications. For a single-tenant application, you can probably predict peak usage times during the day and then schedule resource-hungry batch processing to other times. In a multi-tenant application, and especially those with users from around the globe, there are less likely to be predictable patterns of usage. However, if you have a large number of tenants it may be that variations in resource usage are averaged out.

SCALING WINDOWS AZURE APPLICATIONS WITH WORKER ROLES

Because Windows Azure applications are typically made up of multiple elements such as web and worker roles, tables, blobs, queues, and caches you must consider how to design the application so that each element can support multi-tenancy within the application as a whole, keeping it available and scalable.

You must also consider how best to achieve these goals within the web and worker roles that run your application code. It is possible, though not advisable, to create a large and complex multi-tenant application that has just a single web role (along with any storage that it requires in the cloud). However, you must then ensure that your single web role can handle multiple tenants and be scalable and available. This will almost certainly require complex code that uses multithreading and asynchronous behavior.

Use worker roles to implement asynchronous background processing tasks in your Windows Azure application.

One of the key reasons for using multiple worker role types in your application is to simplify some aspects of the design of your application. For example, by using worker roles you can easily implement background processing tasks, and by using queues you can implement asynchronous behavior. Furthermore, by using multiple role types you can scale each one independently. You might have four instances of your web role, two instances of worker role A, two instances of worker role B, and eight queues. You could also scale roles vertically, for example worker role A could be a small instance, and worker role B a large instance.

By using worker roles to handle storage interactions in your application, and queues to deliver storage insert, update, and delete requests to the worker role, you can implement load leveling. This is particularly important in the Windows Azure environment because both Windows Azure storage and SQL Database can throttle requests when the volume of requests gets too high.

Scalability is an issue for both single-tenant and multi-tenant architectures. Although it may be acceptable to allow certain operations at certain times to utilize most of the available resources in a single-tenant application (for example, calculating aggregate statistics over a large dataset at 2:00 A.M.), this is not an option for most multi-tenant applications where different tenants have different usage patterns.

You can use worker roles in Windows Azure to offload resource-hungry operations from the web roles that handle user interaction. These worker roles can perform tasks asynchronously when the web roles do not require the output from the worker role operations to be immediately available.

The timing of maintenance tasks is typically more difficult to plan for multi-tenant applications. In a single-tenant application there may be windows of time to perform system maintenance without affecting users. This is much less likely to be the case in a multi-tenant application. Chapter 7, "Managing and Monitoring Muli-Tenant Applications," discusses this issue in more detail.

Example Scenarios for Worker Roles

The following table describes some example scenarios where you might partition the functionality of the application into separate worker roles for asynchronous job processing. Not all of these scenarios come from the Surveys application; but, for each scenario, the table specifies how to trigger the job and how many worker role instances it could use.

Scenario	Description	Solution
Update survey statistics	The survey owner wants to view the summary statistics of a survey, such as the total number of responses and average scores for a question. Calculating these statistics is a resource intensive task.	Every time a user submits a survey response the application puts a message in a queue named **statistics-queue** with a pointer to the survey response data. Every ten minutes a worker role retrieves the pending messages from the **statistics-queue** queue and adjusts the survey statistics to reflect those survey responses. Only one worker instance should do the calculation over a queue to avoid any concurrency issues when it updates the statistics table. **Triggered by**: Time **Execution model**: Single worker or multiple workers with concurrency control
Dump survey data to Windows Azure SQL Database	The survey owner wants to analyze the survey data using a relational database. Transferring large volumes of data is a time consuming operation.	The survey owner requests the application export the responses for a survey. This action creates a row in a table named **exports** and puts a message in a queue named **export-queue** pointing to that row. Any worker can dequeue messages from the **export-queue** queue and execute the export. After it finishes, it updates the row in the **exports** table with the status of the export process. **Triggered by**: Message in queue **Execution model**: Multiple workers
Store a survey response	Every time a respondent completes a survey, the response data must be reliably persisted to storage. The user should not have to wait while the application persists the survey data.	When a user submits a survey response the application writes the raw survey data to blob storage and puts a message in a queue named **responses-queue**. A worker role polls the **responses-queue** queue and, when a new message arrives, it stores the survey response data in table storage and puts a message in the **statistics-queue** queue to recalculate the statistics. **Triggered by**: Message in queue **Execution model**: Multiple workers
Heartbeat	Many workers running in a grid-like system have to send a "ping" at a fixed time interval to indicate to a controller that they are still active. The heartbeat message must be sent reliably without interrupting the worker's main task.	Every minute each worker executes a piece of code that sends a "ping." **Triggered by**: Time **Execution model**: Multiple workers

You can scale the "Update survey statistics" scenario described in the preceding table by using one queue and one worker role instance for every tenant, or even for every survey. What's important is that only one worker role instance should process and update data that is mutually exclusive within the dataset.

Looking at these example scenarios suggests you can categorize worker roles that perform background processing according to the criteria in the following table.

Trigger	Execution	Types of tasks
Time	Single worker	An operation on a set of data that changes frequently, and that requires an exclusive lock to avoid concurrency issues. Examples include aggregation, summarization, and denormalization. You may have multiple workers running, but you need some kind of concurrency control to avoid corrupting the data. Depending on the scenario you need to choose between optimistic and pessimistic locking by determining which approach enables the highest throughput.
Time	Multiple workers	An operation on a set of data that is mutually exclusive from other sets so that there are no concurrency issues. Independent operations that don't work over data, such as a "ping."
Message in a queue	Single or multiple workers	An operation on a small number of resources (for example, a blob or several table rows) that should start as soon as possible.

> In the scenario where you use a single worker to update data that requires exclusive access, you may be able to use multiple workers if you can implement a locking mechanism to manage concurrent access. If you implement concurrency control with multiple workers to avoid corrupting shared data, you must choose between optimistic and pessimistic locking by determining which approach enables the highest throughput in your particular scenario.

Triggers for Background Tasks

The trigger for a background task could be a timer or a signal in the form of a message in a queue. Time-based background tasks are appropriate when the task must process a large quantity of data that trickles in little by little. This approach is cheaper and will offer higher throughput than an approach that processes each piece of data as it becomes available because you can batch the operations and reduce the number of storage transactions required to process the data. You can implement a time-based trigger by using a Timer object in a worker role that executes a task at fixed time interval.

*For flexibility in scheduling tasks you could use the Windows Task Scheduler within a worker role, or a specialized library such as **Quartz.NET**.*

You can pull multiple messages from a queue in a single transaction.

If the frequency at which new items of data become available is low and there is a requirement to process the new data as soon as possible, using a message in a queue as a trigger is the appropriate approach. You can implement a message-based trigger in a worker role by creating an infinite loop that polls a message queue for new messages. You can retrieve either a single message or multiple messages from the queue and execute a task to process the message or messages.

Execution Model

In Windows Azure you typically execute background tasks by using worker roles. You could partition the application by having a separate worker role type for each type of background task in your application, but this approach means that you will need at least one separate worker role instance for each type of task. Often you can make better use of the available compute resources by having one worker role handle multiple types of tasks, especially when you have high volumes of data, because this approach reduces the risk of underutilizing your compute nodes. This approach, often referred to as role conflation, involves several trade-offs:

- The first trade-off is the complexity and cost of implementing role conflation against the potential cost savings that result from reducing the number of running worker role instances.

- The second trade-off is the cost savings of running fewer role instances against the flexibility of being able to scale the resources assigned to individual tasks.

- The third trade-off is the time required to implement and test a solution that uses role conflation, and other business priorities such as time-to-market. In this scenario you can still scale out the application by starting up additional instances of the worker role.

Figure 1 shows the two scenarios for running tasks in worker roles.

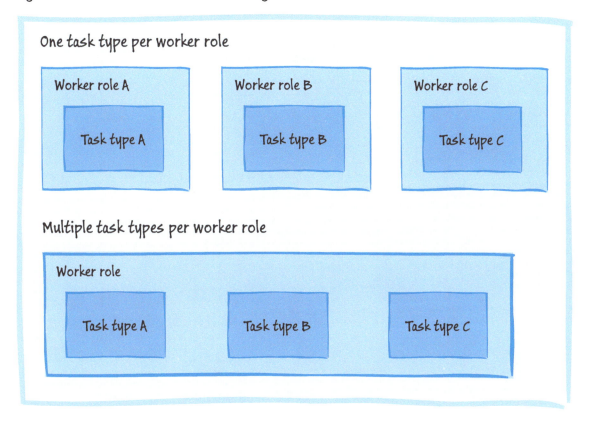

FIGURE 1
Handling multiple background task types

In the scenario where multiple instances of a worker role can all execute the same set of task types you must distinguish between the task types where it is safe to execute the task in multiple worker roles simultaneously, and the task types where it is only safe to execute the task in a single worker role at a time.

To ensure that only one copy of a task can run at a time you must implement a locking mechanism. In Windows Azure you could use a message on a queue or a lease on a blob for this purpose. The diagram in Figure 2 shows that multiple copies of Tasks A and C can run simultaneously, but only one copy of Task B can run at any one time. One copy of Task B acquires a lease on a blob and runs; other copies of Task B will not run until they can acquire the lease on the blob.

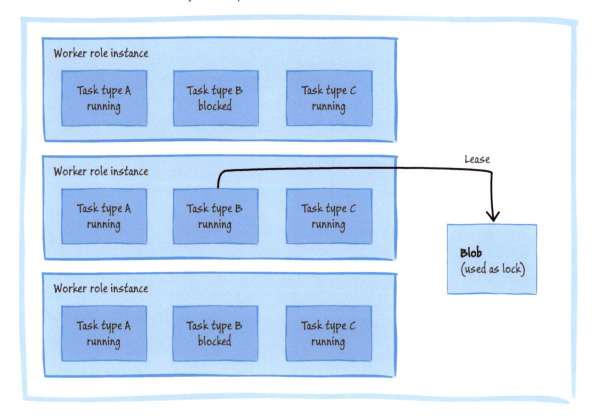

FIGURE 2
Multiple worker role instances

The MapReduce Algorithm

For some Windows Azure applications, being limited to a single task instance for complex long-running calculations may have a significant impact on performance and may limit the scalability of the application. In these circumstances the MapReduce algorithm may provide a way to parallelize the calculations across multiple worker role instances.

The original concepts behind MapReduce come from the **map** and **reduce** functions that are widely used in functional programming languages such as Haskell, F#, and Erlang. In the current context, MapReduce is a programming model that enables you to parallelize operations on a large dataset. In the case of the Surveys application, Tailspin considered using this approach to calculate the summary statistics by using multiple, parallel tasks instead of a single task. The benefit would be to speed up the calculation of the summary statistics by using multiple worker role instances.

For the **Surveys** application, speed is not a critical factor in the calculation of the summary statistics. Tailspin is willing to tolerate a delay while this summary data is calculated, so it does not use **MapReduce**.

> *Hadoop on Windows Azure provides a framework that enables you to optimize the type of operations that benefit from the MapReduce programming model. For more information, see "Introduction to Hadoop on Windows Azure."*

GOALS AND REQUIREMENTS

This section describes the availability, scalability, and elasticity goals and requirements that Tailspin has for the Surveys application.

Performance and Scalability when Saving Survey Response Data

When a user completes a survey, the application must save the user's answers to the survey questions to storage so that the survey creator can access and analyze the results as required. The way that the application saves the summary response data must enable the Surveys application to meet the following three requirements:

- The owner of the survey must be able to browse the results.
- The application must be able to calculate summary statistics from the answers.
- The owner of the survey must be able to export the answers in a format that enables detailed analysis of the results.

Tailspin expects to see a very large number of users completing surveys, and so the process that initially saves the data should be as efficient as possible. The application can handle any processing of the data after it has been saved by using an asynchronous worker process. For information about the design of this background processing functionality in the Surveys application, see the section "Partitioning Web and Worker Roles" in Chapter 4, "Partitioning Multi-Tenant Applications," of this guide.

The focus in this chapter is on the way the Surveys application stores the survey answers. Whatever type of storage the Surveys application uses, it must be able to support the three requirements listed earlier while ensuring the application remains scalable. Storage costs are also a significant factor in the choice of storage type because survey answers account for the majority of the application's storage requirements; both in terms of space used and the number of storage transactions required.

Depending on the volume of survey responses received, transaction costs may become significant because calculating summary statistical data and exporting survey results will require the application to read survey responses from storage.

Calculating summary statistics is an expensive operation if there are a large number of responses to process.

Summary Statistics

Tailspin anticipates that some surveys may have thousands, or even hundreds of thousands of respondents, and wants to make sure that the public website remains responsive for all users at all times. At the same time, survey owners want to be able to view summary statistics calculated from the survey responses submitted to date.

In addition to browsing survey responses, subscribers must be able to view some basic summary statistics that the application calculates for each survey, such as the total number of responses received, histograms of the multiple-choice results, and aggregations such as averages of the range results. The Surveys application provides a predetermined set of summary statistics that cannot be customized by subscribers. Subscribers who want to perform a more sophisticated analysis of their survey responses can export the survey data to a Windows Azure SQL Database instance.

Because of the expected volume of survey response data, Tailspin anticipates that generating the summary statistics will be an expensive operation because of the large number of storage transactions that must occur when the application reads the survey responses. Tailspin wants to have a different SLA for premium and standard subscribers. The Surveys application will prioritize updating the summary statistics for premium subscribers over updating the summary statistics for standard subscribers.

The public site where respondents fill out surveys must always have fast response times when users save their responses, and it must record the responses accurately so that there is no risk of any errors in the data when a subscriber comes to analyze the results.

The developers at Tailspin also want to be able to run comprehensive unit tests on the components that calculate the summary statistics without any dependencies on Windows Azure storage.

Geo-location in the Surveys Application

Tailspin plans to offer subscriptions to the Surveys application to a range of users, from large enterprises to individuals. These subscribers could be based anywhere in the world, and may want to run surveys in other geographic locations. Each subscriber will select a geographic location during the on-boarding process; this location will be where the subscriber creates surveys, accesses the survey response data, and is also the default location for publishing surveys. Windows Azure allows you to select a geographic location for your Windows Azure services so that you can host your application close to your users.

> There are also integration tests that verify the end-to-end behavior of the application using Windows Azure storage.

Tailspin wants to allow subscribers to the Surveys service to override their default geographical location when they publish a survey. By default, a U.S. based subscriber publishes surveys to a U.S. based instance of the Surveys application, and a European subscriber would probably want to choose a Europe based service. However, it's possible that a subscriber might want to run a survey in a different geographic region than the one the subscriber is located in. Figure 3 shows how a U.S. based subscriber might want to run a survey in Europe:

The Surveys application is a "geo-aware" service.

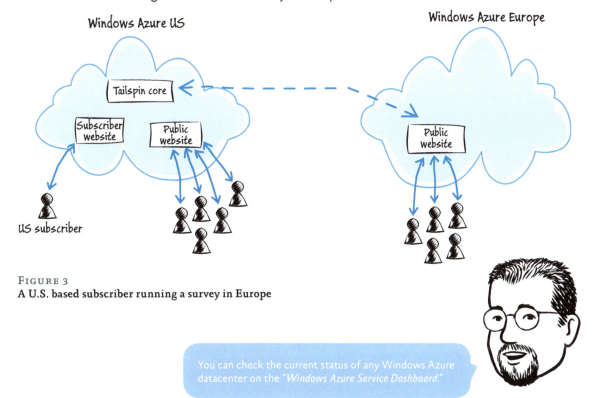

FIGURE 3
A U.S. based subscriber running a survey in Europe

> You can check the current status of any Windows Azure datacenter on the *"Windows Azure Service Dashboard."*

Of course, this doesn't address the question of how users will access the appropriate datacenter. If a survey is hosted in only one datacenter, the subscriber would typically provide a link for users that specifies the survey in that datacenter; for example, *http://eusurveys.tailspin.com/tenant1/europesurvey*. A tenant could also use a CNAME in its DNS configuration to map an address such as *http://eu.tenant1.com/surveys/tenant1/europesurvey* to the actual URL of the survey installed in the North Europe datacenter at *http://eusurveys.tailspin.com/tenant1/europesurvey*.

However, if a subscriber decides to run an international survey and host it in more than one datacenter, Tailspin could allow it to configure a Windows Azure Traffic Manager policy that routes users' requests to the appropriate datacenter—the one that will provide the best response times for their location.

For more information, see the section "Reducing Network Latency for Accessing Cloud Applications with Windows Azure Traffic Manager" in Appendix E of the guide *Building Hybrid Applications in the Cloud on Windows Azure*."

Making the Surveys Application Elastic

In addition to ensuring that Tailspin can scale out the Surveys application to meet higher levels of demand, Tailspin wants the application to be elastic and automatically scale out during anticipated and unexpected increases in demand for resources. The application should also automatically release resources when it no longer needs them in order to control its running costs.

Scalability

In addition to partitioning the application into web and worker roles, queues, and storage, Tailspin plans to investigate any other features of Windows Azure that might enhance the scalability of the application. For example, it will evaluate whether the Surveys application will benefit from using the Content Delivery Network (CDN) to share media resources and offload some of the work performed by the web roles. It will also evaluate whether Shared Access Signatures (SAS) will reduce the workload of worker roles by making blob storage directly and securely available to clients.

Tailspin also wants to be able to test the application's behavior when it encounters high levels of demand. Tailspin wants to verify that the application remains available to all its users, and that the automatic scaling that makes the application elastic performs effectively.

The scalability of the solution can be measured only by stress testing the application. Chapter 7, "Managing and Monitoring Multi-Tenant Applications," outlines the approach that Tailspin took to stress test the Surveys application, and describes some of its findings.

Tailspin expects that elasticity will be important for the public web site and the worker role. However, usage of the private subscriber web site will be much lower and Tailspin does not expect to have to scale this site automatically.

Overview of the Solution

This section describes the approach taken by Tailspin to meet the goals and requirements that relate to making the application available, scalable, and elastic.

Options for Saving Survey Responses

As you saw in Chapter 3 of this guide, Tailspin chose to use Windows Azure blob storage to store survey responses submitted by users filling out surveys in the public survey website. You will see more details in this section of how Tailspin made that decision, and the factors it considered.

In addition to the two options, writing directly to storage and using the delayed write pattern, that are discussed below, Tailspin also considered using shared access signatures to enable the client browser to save survey responses directly to blob storage and post a notification directly to a message queue. The benefit of this approach would be to offload the work of saving survey response data from the web role. However, they discounted this approach because of the complexity of implementing a reliable cross-browser solution and because of the loss of control in the web role over the process of saving survey responses.

Writing Directly to Storage

Figure 4 shows the process Tailspin implemented for saving the survey responses by writing them directly to blob storage using code running in the web role instances.

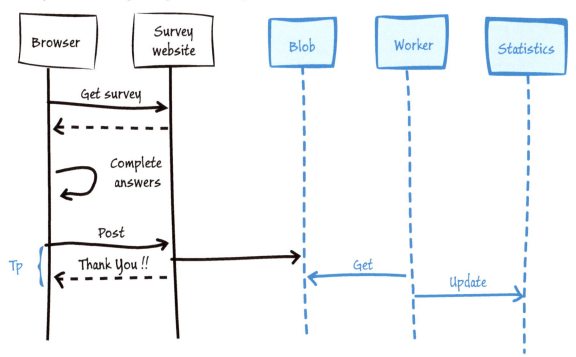

FIGURE 4
Saving survey responses and generating statistics

Figure 4 also shows how the worker role instances collect each new set of responses from storage and uses them to update the summary statistics for that survey. Not shown in this figure is the way that the web role informs the worker role that a new set of answers has been saved in blob storage. It does this by sending a message containing the identifier of the new set of survey answers to a notification queue that the worker role listens on.

Amongst the concerns the developers had when choosing a storage mechanism was that saving a complete set of answers directly to Windows Azure storage from the web role could cause a delay (shown as Tp in Figure 4) at the crucial point when a user has just completed a survey. If a user has to wait while the answers are saved, he or she may decide to leave the site before the operation completes. To address this concern, the developers considered implementing the delayed write pattern.

Using the Delayed Write Pattern

The delayed write pattern is a mechanism that allows code to hand off tasks that may take some time to complete, without needing to wait for them to finish. The tasks can execute asynchronously as background processes, while the code that initiated them continues to other perform other work or returns control to the user.

The delayed write pattern is particularly useful when the tasks that must be carried out can run as background processes, and you want to free the application's UI for other tasks as quickly as possible. However, it does mean that you cannot return the result of the background process to the user within the current request. For example, if you use the delayed write pattern to queue an order placed by a user, you will not be able to include the order number generated by the background process in the page you send back.

In Windows Azure, background tasks are typically initiated by allowing the UI to hand off the task by sending a message to a Windows Azure storage queue. Because queues are the natural way to communicate between the roles in a Windows Azure application, it's tempting to consider using them for an operation such as saving data collected in the UI. The UI code can write the data to a queue and then continue to serve other users without needing to wait for operations on the data to be completed.

Figure 5 shows the delayed write pattern that the Surveys application could use to save the results of a filled out survey to Windows Azure storage.

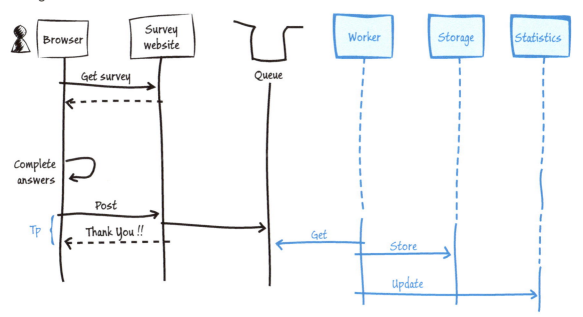

FIGURE 5
Delayed write pattern for saving survey responses in the Surveys application

> Based on tests that Tailspin performed, writing to a queue takes approximately the same time as writing to blob storage, and so there is no additional overhead for the web role compared to saving the data directly to blob storage when using the delayed write pattern.

In this scenario a user browses to a survey, fills it out, and then submits his or her answers to the Surveys website. The code running in the web role instance puts the survey answers into a message on a queue and returns a "Thank you" message to the user as quickly as possible, minimizing the value of Tp in Figure 5. One or more tasks in the worker role instances are then responsible for reading the survey response from the queue, saving it to Windows Azure storage, and updating the summary statistics. This operation must be idempotent to avoid any possibility of double counting and skewing the results.

Surveys is a "geo-aware" application. For example, a European company might want to run a survey in the U.S. but analyze the data locally in Europe; it could use a copy of the Surveys website and queues running in a datacenter in the U.S., and use worker roles and a storage account hosted in a datacenter in Europe. Moving data between data centers will incur bandwidth costs.

Handling Large Messages

There is a 64 kilobyte (KB) maximum size for a message on a Windows Azure queue, or 48 KB when using Base64 encoding for the message, so the approach shown in Figure 5 works only if the size of each survey response is less than the maximum. In most cases, except for very large surveys, it's unlikely that the answers will exceed 48 KB but Tailspin must consider how it will handle this limitation.

One option would be to implement a hard limit on the total response size by limiting the size of each answer, or by checking the total response size using JavaScript code running in the browser. However, Tailspin wants to avoid this as it may limit the attractiveness of its service to some subscribers.

Figure 6 shows how Tailspin could modify the delayed write pattern solution to handle survey results that are greater than 64 KB in size. It includes an optimization by saving messages that are larger than 64 KB to Windows Azure blob storage and placing a message on the "Big Surveys" queue to notify the worker role, which will read these messages from blob storage. Messages that are smaller than 64 KB are placed directly onto a queue as in the previous example.

> When you calculate the size of messages you must consider the effect of any encoding, such as Base64, you use to encode the data before you place it in a message.

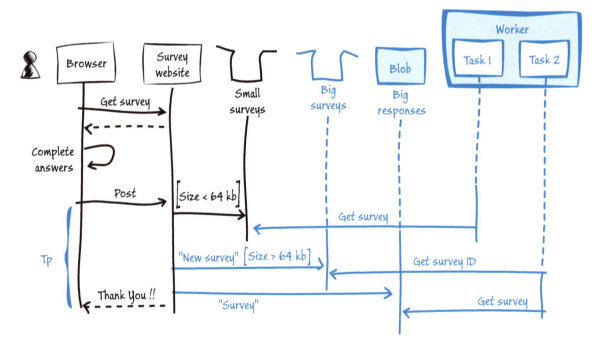

FIGURE 6
Handling survey results greater than 64 KB in size

The worker role now contains two tasks dedicated to saving survey responses and updating the summary statistics:

- Task 1 polls the "Small Surveys" queue and picks up the sets of answers. Then (not shown in the figure) it writes them to storage and updates the summary statistics.
- Task 2 polls the "Big Surveys" queue and picks up messages containing the identifier of the new answers sets that the web role has already written to storage. Then (not shown in the figure) it retrieves the answers from storage and uses them to update the summary statistics.

Notice that, for messages larger than the limit for the queue, the process is almost identical to that described in Figure 4 where Tailspin was not using the delayed write pattern.

> An alternative approach to overcoming the constraint imposed by the maximum message size in Windows Azure queues is to use Windows Azure Service Bus instead. Service Bus queues can handle messages up to 256 KB in size, or 192 KB after Base64 encoding. For more details see "Windows Azure Queues and Windows Azure Service Bus Queues - Compared and Contrasted."

Another variation on the approach described here is to use a single queue that transports two different message types. One message type holds a full survey response as its payload; the other message type holds the address of the blob where the big survey response is stored. You can then implement a **RetrieveSurvey** method in your messaging subsystem that returns either a small or big survey response from the queue to your worker role. Your messaging subsystem now encapsulates all of the logic for handling different response sizes, hiding it from the rest of your application.

Scaling the Worker Role Tasks

In the initial solution Tailspin implemented, writing directly to storage from the web role, the worker role instances had only one task to accomplish: updating the summary statistics. When using the delayed write pattern the worker roles must accomplish two tasks: saving the answers to storage (where the answer set is smaller than the limit for a queue) and updating the summary statistics.

It's possible that Tailspin will want, or need, to scale these two tasks separately. It's vital that new answers are saved to storage as quickly as possible, whereas calculating the summary statistics may not be such an urgent requirement. The summary statistics can be recalculated from the answers should a failure occur, but the converse is not possible. Tailspin also wants to be able to differentiate the service level for premium and standard subscribers by ensuring that summaries for premium subscribers are available more quickly.

To scale the tasks separately Tailspin would need to use two separate worker roles:

- A worker role that just updates the statistics by polling a queue for messages containing the identifier of new answer sets. In Figure 6 this is the "Big Surveys" queue that the web role uses to inform worker roles that it has saved directly to storage a new set of answers that is larger than the limit for a queue.
- A worker role that just saves new answer sets to storage by polling a queue for messages that contain the answers. In Figure 6 this is the "Small Surveys" queue that the web role uses to post sets of answers that are smaller than the limit for a queue to worker roles. However, this worker role would then need to inform the worker role that updates the statistics that it has saved to storage a new set of answers. It would do this by sending a message containing the identifier of the new answer set to the "Big Surveys" queue shown in Figure 6.

To provide different levels of service, such as the speed of processing summary statistics, Tailspin could use separate queues for premium and standard subscribers and configure the worker role that saves the answers to storage to send the notification message to the appropriate queue. The worker role instances that poll these two queues could do so at different rates, or they could use an algorithm that gives precedence to premium subscribers.

Comparing the Options

To identify the best solution for saving survey responses in the Surveys application, the developers at Tailspin considered several factors:

- How to minimize the delay between a user submitting a set of answers and the website returning the "Thank you" page.
- The opportunities for minimizing the storage transaction costs encountered with different approaches for saving the answers and calculating the summary statistics.
- The impact on other parts of the system from the approach they choose for saving the answers.
- The choice of persistent storage mechanism (blobs or tables) that best suits the approach they choose for saving and processing the answers, and will have the least impact on other parts of the system while still meeting all their requirements.

To help them understand the consequences of their choices, Tailspin's developers created the following table to summarize the operations that must be executed for each of the three approaches they considered.

Option	Answer set size	Web role storage transactions	Worker role storage transactions	Total # of transactions
Write answers directly to storage from the web role.	Any	Save answers to storage. Post message to notification queue.	Read message from notification queue. Read answers from storage. Read current summary statistics. Write updated summary statistics. Call complete on notification queue.	Seven
Use the delayed write pattern with the worker role handling the tasks of writing to storage and calculating summary statistics.	< 64 KB	Post answers to "Small Surveys" queue.	Read answers from "Small Surveys" queue. Write answers to storage. Read current summary statistics. Write updated summary statistics. Call complete on "Small Surveys" queue.	Six
	> 64 KB	Save answers to storage. Post message to "Big Surveys" queue.	Read message from "Big Surveys" queue. Read answers from storage. Read current summary statistics. Write updated summary statistics. Call complete on "Big Surveys" queue.	Seven
Use the delayed write pattern with separate worker roles for the tasks of writing to storage and calculating summary statistics.	< 64 KB	Post answers to "Small Surveys" queue.	Save survey worker role: Read answers from "Small Surveys" queue. Write answers to storage. Call complete on "Small Surveys" queue. Post message to "Big Surveys" queue. Update statistics worker role: Read message from "Big Surveys" queue. Read answers from storage. Read current summary statistics. Write updated summary statistics. Call complete on "Big Surveys" queue.	Ten
	> 64 KB	Save answers to storage. Post message to "Big Surveys" queue.	Update statistics worker role: Read message from "Big Surveys" queue. Read answers from storage. Read current summary statistics. Write updated summary statistics. Call complete on "Big Surveys" queue.	Seven

Some points to note about the contents of the table are:

• Worker roles can read messages from a queue in batches, which reduces the storage transaction costs because reading a batch of messages counts as a single transaction. However, this means that there may be a delay between answers being submitted and the worker role processing them and, when using the delayed write pattern with small answer sets, saving them to storage.

- Using the delayed write pattern with two separate worker role types allows you to scale the two tasks (writing to storage and calculating the summary statistics) separately. This means that the two tasks must access the answers separately and in the correct order. One task reads them from the answers queue, writes them to storage, and only then posts a message to a queue to indicate new answers are available. The second task reads the answers from storage when the message is received, and updates the statistics.

- Using the delayed write pattern when messages are larger than the limit for the queue is not really the delayed write pattern at all. It is fundamentally the same as the original approach of saving the answers direct to storage.

- Because the majority of answer sets are likely to be smaller than the limit for the queue, the third option that uses separate worker role types will typically use more storage transactions than if there was a predominance of large answer sets.

Keeping the UI Responsive when Saving Survey Responses

A key design goal is to minimize the time it takes to save a survey response and return control to the UI. Tailspin does not want survey respondents to leave the site while they wait for the application to save their survey responses. Irrespective of the way that the survey responses are saved to storage, the Surveys application will use a task in the worker role instances to calculate and save the summary statistics in the background after the responses are saved.

The initial approach that Tailspin implemented in the Surveys application requires the web role to perform two operations for each set of answers that users submit. It must first save them to storage and then, if that operation succeeds, post a message to the notification queue so that worker roles know there is a new survey response available.

When using the delayed write pattern and the total size of the answer set is smaller than the limit for Windows Azure storage queues, the web role instances need to perform only one operation. They just need to post the answers to a queue, and all of the processing will occur in the background. The worker roles will write the answers to storage and update the summary statistics; meanwhile the web role can return the "Thank you" page immediately.

If the total size of the answer set is larger than the limit for Windows Azure storage queues, the web role instances will need to perform two operations: saving the answers and then sending a message to the notification queue. However, it is expected that the vast majority of surveys will not produce answer sets that are larger than the limit for a queue.

Even if Tailspin wants to offer premium subscribers the capability for their summary statistics to be updated more quickly than those of standard subscribers, and does this by using two separate worker role types, the web role will still need to perform only one operation unless the answers set size is larger than the limit for a queue.

Therefore, the most efficient option from the point of view of minimizing UI delay will be to use the delayed write pattern because, in the vast majority of cases, it will require only a single operation within the web role code.

Minimizing the Number of Storage Transactions

Reading and writing survey responses account for the majority of storage transactions in the Tailspin Surveys application, and with high monthly volumes this can account for a significant proportion of Tailspin's monthly running costs.

The option that requires the least number of storage transactions is the delayed write pattern with the worker role saving the answers and calculating the summary as one operation. This option will require an additional storage transaction for survey answers larger than the limit for a queue, but this is not expected to occur very often. However, as you saw in the previous section, this option limits the capability to scale the tasks separately in the worker role, and may make using separate queues for premium and standard subscribers more complicated.

The next best option is to write the answers directly to storage using code in the web role. To save a complete survey response directly to blob storage requires a single storage transaction. If the Surveys application used Windows Azure table storage instead of blob storage, and can use a single entity group transaction to save a survey answers to table storage, it could also save each complete survey response in a single transaction.

The third option, using the delayed write pattern with separate worker role types for saving the answers and updating the summary statistics will require the highest number of storage transactions for the vast majority of survey answers.

The Impact on Other Parts of the System

The decision on the type of storage to use (blob or table) and whether to use the delayed write pattern can have an impact on other parts of the application, and on the associated systems and services. Tailspin's developers carried out a set of spikes to determine whether using blob storage would make it difficult or inefficient to implement the other parts of the Surveys application that read the survey response data. This includes factors such as paging through survey responses in the UI, generating summary statistics, and exporting to a SQL Database instance.

They determined that using blob storage for storing survey response data will not introduce any significant additional complexity to the implementation, and will not result a significant increase in the number of storage transactions within the system. Chapter 3, "Choosing a Multi-Tenant Data Architecture," describes how Tailspin implemented both paging through survey responses stored in blob storage and exporting survey response data to SQL Database. The section *"Options for Generating Summary Statistics"* in this chapter describes how Tailspin implemented the export feature in the Surveys application.

To be able to save a complete survey response in a single entity group transaction, the survey answer set must have fewer than 100 answers, and must be stored in a single table partition. An entity group transaction batches a group of changes to a single table partition into a single, atomic operation that counts as a single storage transaction. An entity group transaction must update fewer than 100 entities and the total request size must be less than 4 MB in size.

Pre-processing the data before the application saves it is typically used to avoid the need to perform the processing every time the data is read. If the application writes the data once, but reads it n times, the processing is performed only once, and not n times.

The delayed write pattern has the advantage that it makes it easy to perform any additional processing on a survey response before it is saved, without delaying the UI. This processing might include formatting the data or adding contextual information. The web role places the raw survey response in a message. The worker role retrieves the message from the queue, performs any required processing on the data in the message, and then saves the processed survey response.

Tailspin did not identify any additional processing that the Surveys application could perform on the survey responses that would help to optimize the processes that read the survey data. The developers at Tailspin determined that they could implement all of these processes efficiently, whether the survey response data was stored in blob or table storage.

Choosing between Blob and Table Storage

The initial assumption of Tailspin's developers during the early design process for the Surveys application was that it should save each survey response as a set of rows in Windows Azure table storage. However, before making the final decision, the developers carried out some tests to find the comparable speed of writing to table storage and blob storage. They created some realistic spikes to compare how long it takes to serialize and save a survey response to a blob with how long it takes to save the same survey response as a set of entities to table storage in a single entity group transaction. They found that, in their particular scenario, saving to blob storage is significantly faster.

When using the delayed write pattern, the additional time to save the survey response data will affect only the worker role. The web role UI code will need only to write the survey responses to a queue. There will be no additional delay for users when submitting their answers. However, the added overhead in the worker role may require extra resources such as additional instances, which will increase the running cost of the application.

If Tailspin chose not to use the delayed write pattern, the increase in time for the web role to write to table storage will have an impact on the responsiveness of the UI. Using table storage will also have an impact when the delayed write pattern is used and the answer sets are predominantly larger than the limit for a queue. Therefore, in order to allow for this possibility and to make future extensions to the application that may require larger messages to be accepted easier, Tailspin chose to store the answers in blob storage.

If you use table storage you must consider how your choice of partition key affects the scalability of your solution both when writing and reading data. If we chose to store the survey answers in table storage we'd need to choose a partition key that allows the Surveys application to save each survey response using an entity group transaction, and read survey responses efficiently when it calculates summary statistics or exports data to SQL Database.

Options for Generating Summary Statistics

To meet the requirements for generating summary statistics, the developers at Tailspin decided to use a worker role to handle the task of generating these from the survey results. Using a worker role enables the application to perform this resource intensive process as a background task, ensuring that the web role responsible for collecting survey answers is not blocked while the application calculates the summary statistics.

Based on the framework for worker roles described in Chapter 4, "Partitioning Multi-Tenant Applications," this asynchronous task is one that will be triggered on a schedule. In addition, because it updates a single set of results, it must run as a single instance process or include a way to manage concurrent access to each set of summary data.

To calculate the survey statistics, Tailspin considered two basic approaches. The first approach is for the task in the worker role to retrieve all the survey responses to date, recalculate the summary statistics, and then save the summary data over the top of the existing summary data. The second approach is for the task in the worker role to retrieve all the survey response data that the application has saved since the last time the task ran, and use this data to adjust the summary statistics to reflect the new survey results.

> *You can use a queue to maintain a list of all new survey responses. The summarization task is triggered on a schedule that determines how often the task should look at the queue for new survey results to process.*

The first approach is the simplest to implement, because the second approach requires a mechanism for tracking which survey results are new. The second approach also depends on it being possible to calculate the new summary data from the old summary data and the new survey results, without re-reading all the original survey results.

For many types of summary statistic (such as total, average, count, and standard deviation) it is possible to calculate the new values based on the current values and the new results. For example if you have already received five answers to a numeric question and you know that the average of those answers is four, then if you receive a new response with an answer of 22, then the new average is $((5 * 4) + 22)/6$ which equals seven. Note that you need to know both the current average and the current number of answers to calculate the new average. However, suppose you want one of your pieces of summary data to be a list of the ten most popular words used in answering a free-text question. In this case, you would always have to process all of the survey answers, unless you also maintained a separate list of all the words used and a count of how often they appeared. This adds to the complexity of the second approach.

The key difference between the two approaches is in the number of storage transactions required to perform the summary calculations: this directly affects both the cost of each approach and time it takes to perform the calculations. The graph in Figure 7 shows the result of an analysis that compares the number of transactions per month of the two approaches for three different daily volumes of survey answers. The graph shows the first approach on the upper line with the Recalculate label, and the second approach on the lower line with the Merge label.

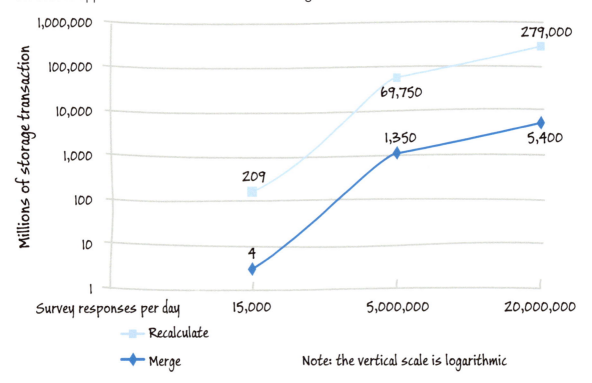

FIGURE 7
Comparison of transaction numbers for alternative approaches to calculating summary statistics

The graph clearly shows that fewer storage transactions are required if Tailspin adopts the merge approach. Tailspin decided to implement the merge approach in the Surveys application.

The vertical cost scale on the chart is logarithmic. The analysis behind this chart makes a number of "worst case" assumptions about the way the application processes the survey results. The chart is intended to illustrate the relative difference in transaction numbers between the two approaches; it is not intended to show absolute numbers.

It is possible to optimize the recalculate approach if you decide to sample the survey answers instead of processing every single one when you calculate the summary data. You would need to perform some detailed statistical analysis to determine what proportion of results you need to select to calculate the summary statistics within an acceptable margin of error.

Scaling out the Generate Summary Statistics Task

The Tailspin Surveys application must be able to scale out to handle an increase in the number of survey respondents. This should include enabling multiple instances of the worker role that performs the summary statistics calculation and builds the ordered list of survey responses. For each survey there is just a single set of summary statistics, so the application must be able to handle concurrent access to a single blob from multiple worker roles without corrupting the data. Tailspin considered four options for handling concurrency:

- Use a single instance of the worker role. While it is possible to scale up by using a larger instance, there is a limit to the scalability of this approach. Furthermore, this option does not include any redundancy if that instance fails.

- Use the MapReduce programming model. This approach would enable Tailspin to use multiple task instances, but would add to the complexity of the solution.

- Use pessimistic concurrency. In this approach, the statistics associated with several specific surveys are locked while a worker role processes a batch of new responses. The worker role reads a batch of messages from the queue, identifies the surveys they are associated with, locks those specific sets of summary statistics, calculates and saves the new summary statistics, and then releases the locks. This would mean that other worker instances trying to update any of the same sets of summary statistics are blocked until the first instance releases the locks.

- Use optimistic concurrency. In this approach, when the worker role instance processes a batch of messages, it checks for each message whether or not another task is updating that specific survey's summary statistics. If another task is already updating the statistics, the current task puts the message back on the queue to be reprocessed later; otherwise it goes ahead with the update.

Tailspin performed stress testing to determine the optimum solution and chose the fourth option—using optimistic concurrency. It allows Tailspin to scale out the worker role instance, allows for a higher throughput of messages than the pessimistic concurrency approach, and offers better performance because it does not require any locking mechanism. Although MapReduce would also work, it adds more complexity to the system than using the optimistic concurrency approach.

> *For more information about the stress tests Tailspin carried out, see Chapter 7, "Managing and Monitoring Multi-tenant Applications." For a description of the MapReduce programming model see the section "The MapReduce Algorithm" earlier in this chapter.*

Using Windows Azure Caching

Chapter 4, *"Partitioning Multi-Tenant Applications,"* describes how the Tailspin uses Windows Azure Caching to support the Windows Azure Caching session state provider, and how Tailspin ensures tenant data is isolated within the cache. Tailspin uses Windows Azure Caching to cache survey definitions and tenant data in order to reduce latency in the public Surveys website.

Tailspin chose to use Windows Azure Caching, configured a cache that is co-located in the Tailspin. Web worker role, and uses 30% of the available memory. Tailspin will monitor cache utilization levels and the performance of the Tailspin.Web role in order to review whether these settings provide enough cache space without affecting the usability of the private tenant web site.

The section "Caching Frequently Used Data" in Chapter 4 shows how caching is implemented in the data access layer. Tailspin Surveys implements caching behavior in the **SurveyStore** and **TenantStore** classes.

Using the Content Delivery Network

This section looks at how the Windows Azure Content Delivery Network (CDN) can improve the user experience. The CDN allows you to cache blob content at strategic locations around the world to make that content available with the maximum possible bandwidth to users, and minimize network latency. The CDN is designed to be used with blob content that is relatively static.

For the Surveys application, the developers at Tailspin have identified two scenarios where they could use the CDN:

• Tailspin is planning to commission a set of training videos with titles such as "Getting Started with the Surveys Application," "Designing Great Surveys," and "Analyzing your Survey Results."

• Hosting the custom images and style sheets that subscribers upload.

In both of these scenarios, users will access the content many times before it's updated. The training videos are likely to be refreshed only when the application undergoes a major upgrade, and Tailspin expects subscribers to upload corporate logos and style sheets that reflect corporate branding. Both of these scenarios will also account for a significant amount of bandwidth used by the application. Online videos will require sufficient bandwidth to ensure good playback quality, and every request to fill out a survey will result in a request for a custom image and style sheet.

One of the requirements for using the CDN is that the content must be in a blob container that you configure for public, anonymous access. Again, in both of the scenarios, the content is suitable for unrestricted access.

For information about the current pricing for the CDN, see the "Caching" section of the page *"Pricing Details"* on the Windows Azure website.

For data cached on the CDN, you are charged for outbound transfers based on the amount of bandwidth you use and the number of transactions. You are also charged at the standard Windows Azure rates for the transfers that move data from blob storage to the CDN. Therefore, it makes sense to use the CDN for relatively static content. With highly dynamic content you could, in effect, pay double because each request for data from the CDN triggers a request for the latest data from blob storage.

To use the CDN with the Surveys application, Tailspin will have to make a number of changes to the application. The following sections describe the solution that Tailspin plans to implement in the future; the current version of the Surveys application does not use the CDN.

Setting the Access Control for the BLOB Containers

Any blob data that you want to host on the CDN must be in a blob container with permissions set to allow full public read access. You can set this option when you create the container by calling the **BeginCreate** method of the **CloudBlobContainer** class or by calling the **SetPermissions** method on an existing container. The following code shows an example of how to set the permissions for a container.

```C#
protected void SetContainerPermissions(String containerName)
{
  CloudStorageAccount cloudStorageAccount =
    CloudStorageAccount.Parse(
      RoleEnvironment.GetConfigurationSettingValue(
      "DataConnectionString "));

  CloudBlobClient cloudBlobClient =
    cloudStorageAccount.CreateCloudBlobClient();

  CloudBlobContainer cloudBlobContainer =
    new CloudBlobContainer(containerName, cloudBlobClient);

  BlobContainerPermissions blobContainerPermissions =
    new BlobContainerPermissions();

  blobContainerPermissions.PublicAccess =
    BlobContainerPublicAccessType.Container;

  cloudBlobContainer.SetPermissions(
                blobContainerPermissions);
}
```

Notice that the permission type used to set full public access is **BlobContainerPublicAccessType. Container**.

Configuring the CDN and Storing the Content

You configure the CDN at the level of a Windows Azure storage account through the Windows Azure Management Portal. After you enable CDN delivery for a storage account, any data in public blob containers is available for delivery by the CDN.

The application must place all the content to be hosted on the CDN into blobs in the appropriate containers. In the Surveys application, media files, custom images, and style sheets can all be stored in these blobs. For example, if a training video is packaged with a player application in the form of some HTML files and scripts, all of these related files can be stored as blobs in the same container.

You must be careful if scripts or HTML files contain relative paths to other files in the same blob container because the path names will be case sensitive. This is because there is no real folder structure within a blob container, and any "folder names" are just a part of the file name in a single, flat namespace.

Configuring URLs to Access the Content

Windows Azure allocates URLs to access blob data based on the account name and the container name. For example, if Tailspin created a public container named "video" for hosting their training videos, you could access the "Getting Started with the Surveys Application" video directly in Windows Azure blob storage at *http://tailspin.blob.core.windows.net/video/gettingstarted.html*. This assumes that the gettingstarted.html page is a player for the media content.

The CDN provides access to hosted content using a URL in the form *http://<uid>.vo.msecnd.net/*, so the Surveys training video would be available on the CDN at *http://<uid>.vo.msecnd.net/video/gettingstarted.html*. Figure 8 illustrates this relationship between the CDN and blob storage.

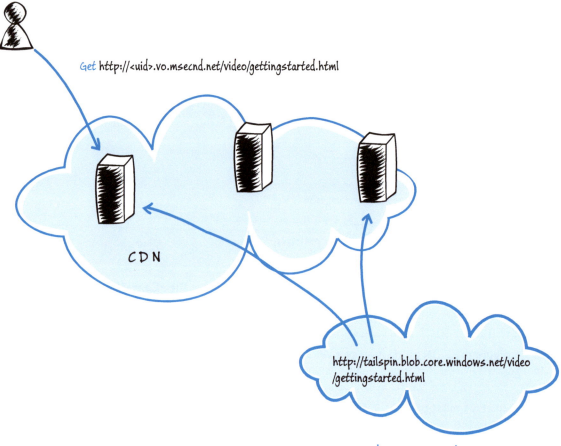

Get http://<uid>.vo.msecnd.net/video/gettingstarted.html

CDN

http://tailspin.blob.core.windows.net/video/gettingstarted.html

Windows Azure Blob Service

FIGURE 8
The Content Delivery Network

You can configure a CNAME entry in DNS to map a custom URL to the CDN URL. For example, Tailspin might create a CNAME entry to make *http://files.tailspin.com/video/gettingstarted.html* point to the video hosted on the CDN. You should verify that your DNS provider configures the DNS resolution to behave efficiently; the performance benefits of using the CDN could be offset if the name resolution of your custom URL involves multiple hops to a DNS authority in a different geographic region.

> *For information about how to use a custom DNS name with your CDN content, see "How to Map CDN Content to a Custom Domain."*

When a user requests content from the CDN, Windows Azure automatically routes their request to the closest available CDN endpoint. If the blob data is found at that endpoint it's returned to the user. If the blob data is not found at the endpoint it's automatically retrieved from blob storage before being returned to the user and cached at the endpoint for future requests.

Setting the Caching Policy

All blobs cached by the CDN have a time-to-live (TTL) period that determines how long they will remain in the cache before the CDN goes back to blob storage to check for updated data. The default CDN caching policy uses an algorithm to calculate the TTL for cached content based on when the content was last modified in blob storage. The longer the content has remained unchanged in blob storage, the greater the TTL, up to a maximum of 72 hours.

> If the blob data is not found at the endpoint, you will incur Windows Azure storage transaction charges when the CDN retrieves the data from blob storage.

> *The CDN retrieves content from blob storage only if it is not in the endpoint's cache, or if it has changed in blob storage.*

You can also explicitly set the TTL by using the **CacheControl** property of the **BlobProperties** class. The following code example shows how to set the TTL to two hours.

```C#
blob.Properties.CacheControl = "max-age=7200";
```

For more information about how to manage expiration policies with CDN, see *"How to Manage Expiration of Blob Content."*

Hosting Tailspin Surveys in Multiple Locations

Hosting a survey in a web role in a different geographic location doesn't, by itself, mean that people filling out the survey will see the best response times when they use the site. To render the survey, the application must retrieve the survey definition from storage, and the application must save the completed survey results to storage. If the application storage is in the U.S. datacenter, there is little benefit to European users accessing a website hosted in the European datacenter.

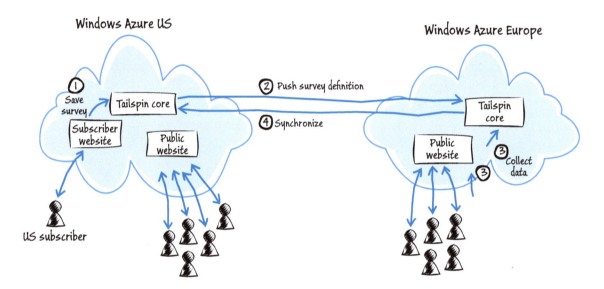

Figure 9 shows how Tailspin designed the application to handle this scenario and resolve the issue just described.

FIGURE 9
Hosting a survey in a different geographic location

The following describes the steps illustrated in Figure 9:

1. The subscriber designs the survey, and the application saves the definition in storage hosted in the U.S. datacenter.

2. The Surveys application pushes the survey definition to another application instance in a European datacenter. This needs to happen only once.

3. Survey respondents in Europe fill out the survey, and the application saves the data to storage hosted in the European datacenter.

4. The application transfers the survey results data back to storage in the U.S. datacenter, where it is available to the subscriber for analysis.

Tailspin must create separate cloud services and storage accounts for each region where it plans to allow subscribers to host surveys.

Tailspin could use caching to avoid the requirement to transfer the survey definitions between data centers in step 2. It could cache the survey definition in Europe and load the survey definition into the cache from the U.S. storage account. This approach means that the Surveys application hosted in Europe must be able to reach the storage account in the U.S. datacenter to be able to load survey definitions into the cache. Geo-replication of data in the U.S. datacenter provides resilience in the case that a major disaster affects the U.S. datacenter, but does not provide resilience in the case of connectivity problems between Europe and the U.S. For more information, see *"Introducing Geo-replication for Windows Azure Storage."*

Synchronizing Survey Statistics

While the application data (the survey definitions and the answers submitted by users) is initially stored in the datacenter where subscriber chose to host the survey, the application copies the data to the data center where the subscriber's account is hosted. Figure 10 illustrates the roles and storage elements in a scenario where the subscriber is based in Europe and has chosen to host a survey in a U.S. datacenter.

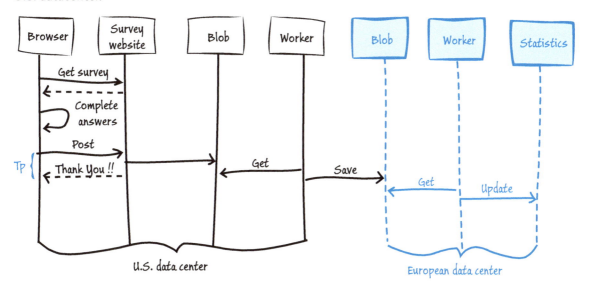

FIGURE 10
Saving the responses from a survey hosted in a different datacenter

This scenario is similar to the scenario described earlier in the section "Writing Directly to Storage" earlier in this chapter, but there is now an additional worker role. This worker role is responsible for moving the survey response data from the datacenter where the subscriber chose to host the survey to the datacenter hosting the subscriber's account. This way, the application transfers the survey data between datacenters only once instead of every time the application needs to read it; this minimizes the costs associated with this scenario.

In some scenarios, it may make sense to pre-process or summarize the data in the datacenter where it's collected and transfer back only the summarized data to reduce bandwidth costs. For the Surveys application, Tailspin decided to move all the data back to the subscriber's datacenter. This simplifies the implementation, helps to optimize the paging feature, ensures that each response is moved between data-centers only once, and ensures that the subscriber has access to all the survey data in the local data center.

The sample application does not currently implement this scenario.

When you deploy a Windows Azure application, you select the sub-region where you want to host the application. This sub-region effectively identifies the datacenter hosting your application. You can also define *affinity groups* to ensure that interdependent Windows Azure applications and storage accounts are grouped together. This improves performance because Windows Azure co-locates members of the affinity group in the same datacenter cluster, and reduces costs because data transfers within the same datacenter do not incur bandwidth charges. Affinity groups offer a small advantage over simply selecting the same sub-region for your hosted services because Windows Azure makes a "best effort" to optimize the location of those services.

Autoscaling and Tailspin Surveys

Tailspin plans to use the Autoscaling Application Block to make the Surveys application elastic. It will configure rules that set the minimum and maximum number of instances for each role type within Tailspin Surveys. For each role type, the minimum number of instances will be five, and Tailspin will adjust the maximum as more subscribers sign up for the service.

Tailspin will also configure dynamic scaling that is based on monitoring key metrics in the Tailspin Surveys application. Initially, it will create rules based on the CPU usage of the different roles and the length of the Windows Azure queues that pass messages to the worker role instances.

Tailspin does not plan to use scheduled rules that adjust the number of role instances based on time and date. However, it will analyze usage of the application to determine whether there are any usage patterns that it can identify in order to preemptively scale the application at certain times.

> *For more information about how to add the Autoscaling Application Block to a Windows Azure application and how to configure your autoscaling rules, see the "Enterprise Library 5.0 Integration Pack for Windows Azure."*

Inside the Implementation

Now is a good time to walk through some of the code in the Tailspin Surveys application in more detail. As you go through this section you may want to download the Visual Studio solution for the Tailspin Surveys application from *http://wag.codeplex.com/*.

The Tailspin Surveys application uses a single worker role type to host two different asynchronous background tasks:

- Calculating summary statistics.
- Exporting survey response data to SQL Database.

The task that calculates the summary statistics also maintains the list of survey responses that enables subscribers to page through responses. Chapter 3, "Choosing a Multi-Tenant Data Architecture," describes this part of the task. Chapter 3 also describes how the export to Windows Azure SQL Database works.

Saving the Survey Response Data Asynchronously

Before the task in the worker role can calculate the summary statistics, the application must save the survey response data to blob storage. The following code from the **SurveysController** class in the TailSpin.Web.Survey.Public project shows how the application saves the survey responses.

```csharp
C#
[HttpPost]
public ActionResult Display(string tenant,
                            string surveySlug,
                            SurveyAnswer contentModel)
{
  var surveyAnswer = CallGetSurveyAndCreateSurveyAnswer(
    this.surveyStore, tenant, surveySlug);

  ...

  for (int i = 0;
       i < surveyAnswer.QuestionAnswers.Count; i++)
  {
    surveyAnswer.QuestionAnswers[i].Answer =
      contentModel.QuestionAnswers[i].Answer;
  }

  if (!this.ModelState.IsValid)
  {
    var model =
      new TenantPageViewData<SurveyAnswer>(surveyAnswer);
    model.Title = surveyAnswer.Title;
    return this.View(model);
  }

  this.surveyAnswerStore.SaveSurveyAnswer(surveyAnswer);

  return this.RedirectToAction("ThankYou");
}
```

The **surveyAnswerStore** variable holds a reference to an instance of the **SurveyAnswerStore** type. The application uses the Unity Application Block (Unity) to initialize this instance with the correct **IAzureBlob** and **IAzureQueue** instances.

> *Unity is a lightweight, extensible dependency injection container that supports interception, constructor injection, property injection, and method call injection. You can use Unity in a variety of ways to help decouple the components of your applications, to maximize coherence in components, and to simplify design, implementation, testing, and administration of these applications. For more information, and to download the application block, see "Unity Application Block."*

The blob container stores the answers to the survey questions, and the queue maintains a list of new survey answers that haven't yet been included in the summary statistics or the list of survey answers.

The **SaveSurveyAnswer** method writes the survey response data to the blob storage and puts a message onto a queue. The action method then immediately returns a "Thank you" message. The following code example shows the **SaveSurveyAnswer** method in the **SurveyAnswerStore** class.

```C#
public void SaveSurveyAnswer(SurveyAnswer surveyAnswer)
{
  var tenant = this.tenantStore
    .GetTenant(surveyAnswer.Tenant);
  if (tenant != null)
  {
    var surveyAnswerBlobContainer = this
      .surveyAnswerContainerFactory
      .Create(surveyAnswer.Tenant, surveyAnswer.SlugName);

    surveyAnswer.CreatedOn = DateTime.UtcNow;
    var blobId = Guid.NewGuid().ToString();
    surveyAnswerBlobContainer.Save(blobId, surveyAnswer);

    (SubscriptionKind.Premium.Equals(
        tenant.SubscriptionKind)
      ? this.premiumSurveyAnswerStoredQueue
      : this.standardSurveyAnswerStoredQueue)
      .AddMessage(new SurveyAnswerStoredMessage
      {
        SurveyAnswerBlobId = blobId,
        Tenant = surveyAnswer.Tenant,
        SurveySlugName = surveyAnswer.SlugName
      });
  }
}
```

Make sure that the storage connection strings in your deployment point to storage in the deployment's datacenter. The application should use local queues and blob storage to minimize latency. Also ensure that you call the CreateIfNotExist method of a queue or blob only once in your storage class constructor, and not in every call to store data. Repeated calls to the CreateIfNotExist method will hurt performance.

This method first locates the blob container for the survey responses. It then creates a unique blob ID by using a GUID, and saves the blob to the survey container. Finally, it adds a message to a queue. The application uses two queues, one for premium subscribers and one for standard subscribers, to track new survey responses that must be included in the summary statistics and the list of responses for paging through answers.

Calculating the Summary Statistics

The team at Tailspin decided to implement the asynchronous background task that calculates the summary statistics from the survey results by using a merge approach. Each time the task runs it processes the survey responses that the application has received since the last time the task ran. It calculates the new summary statistics by merging the new results with the old statistics.

Worker role instances, defined in the TailSpin.Workers.Surveys project, periodically scan two queues for pending survey answers to process. One queue contains a list of unprocessed responses to premium subscribers' surveys; the other queue contains the list of unprocessed responses to standard subscribers' surveys.

The worker role instances executing this task use an optimistic concurrency approach when they try to save the new summary statistics. If one instance detects that another instance updated the statistics for a particular survey while it was processing a batch of messages, it does not perform the update for this survey and puts the messages associated with it back onto the queue for processing again.

The following code example from the **UpdatingSurveyResults-SummaryCommand** class shows how the worker role processes each temporary survey answer and then uses them to recalculate the summary statistics.

```csharp
C#
public class UpdatingSurveyResultsSummaryCommand :
                IBatchCommand<SurveyAnswerStoredMessage>
{
  private readonly
    IDictionary<string, TenantSurveyProcessingInfo>
    tenantSurveyProcessingInfoCache;
  private readonly ISurveyAnswerStore surveyAnswerStore;
  private readonly
    ISurveyAnswersSummaryStore surveyAnswersSummaryStore;
```

It's possible that the role could fail after it adds the survey data to blob storage but before it adds the message to the queue. In this case, the response data would not be included in the summary statistics or the list of responses used for paging. However, the response would be included if the user exported the survey to Windows Azure SQL Database. Tailspin has decided that this is an acceptable risk in the Surveys application.

```
public UpdatingSurveyResultsSummaryCommand(
  IDictionary<string, TenantSurveyProcessingInfo>
    processingInfoCache,
  ISurveyAnswerStore surveyAnswerStore,
  ISurveyAnswersSummaryStore surveyAnswersSummaryStore)
{
  this.tenantSurveyProcessingInfoCache =
                processingInfoCache;
  this.surveyAnswerStore = surveyAnswerStore;
  this.surveyAnswersSummaryStore =
                surveyAnswersSummaryStore;
}

public void PreRun()
{
  this.tenantSurveyProcessingInfoCache.Clear();
}

public bool Run(SurveyAnswerStoredMessage message)
{
  if (!message.AppendedToAnswers)
  {
    this.surveyAnswerStore
      .AppendSurveyAnswerIdToAnswersList(
        message.Tenant,
        message.SurveySlugName,
        message.SurveyAnswerBlobId);
    message.AppendedToAnswers = true;
    message.UpdateQueueMessage();
  }

  var surveyAnswer = this.surveyAnswerStore
    .GetSurveyAnswer(
      message.Tenant,
      message.SurveySlugName,
      message.SurveyAnswerBlobId);

  var keyInCache = string.Format(
    CultureInfo.InvariantCulture, "{0}-{1}",
    message.Tenant, message.SurveySlugName);
  TenantSurveyProcessingInfo surveyInfo;

  if (!this.tenantSurveyProcessingInfoCache
      .ContainsKey(keyInCache))
  {
    surveyInfo = new TenantSurveyProcessingInfo(
              message.Tenant, message.SurveySlugName);
    this.tenantSurveyProcessingInfoCache[keyInCache] =
              surveyInfo;
  }
```

```
    else
    {
      surveyInfo =
        this.tenantSurveyProcessingInfoCache[keyInCache];
    }

    surveyInfo.AnswersSummary.AddNewAnswer(surveyAnswer);
    surveyInfo.AnswersMessages.Add(message);

    return false; // Don't remove the message from the queue
  }
  public void PostRun()
  {
    foreach (var surveyInfo in
             this.tenantSurveyProcessingInfoCache.Values)
    {
      try
      {
        this.surveyAnswersSummaryStore
          .MergeSurveyAnswersSummary(
            surveyInfo.AnswersSummary);

        foreach (var message in surveyInfo.AnswersMessages)
        {
          try
          {
            message.DeleteQueueMessage();
          }
          catch (Exception e)
          {
            TraceHelper.TraceWarning(
              "Error deleting message for '{0-1}': {2}",
              message.Tenant, message.SurveySlugName,
              e.Message);
          }
        }
      }
      catch (Exception e)
      {
        // Do nothing. This leaves the messages in
        // the queue ready for processing next time.
        TraceHelper.TraceWarning(e.Message);
      }
    }
  }
}
```

The Surveys application uses Unity to initialize an instance of the **UpdatingSurveyResultsSummaryCommand** class, and the **surveyAnswerStore** and **surveyAnswersSummaryStore** variables. The **surveyAnswerStore** variable is an instance of the **SurveyAnswerStore** type that the **Run** method uses to read the survey responses from blob storage.

The **surveyAnswersSummaryStore** variable is an instance of the **SurveyAnswersSummary** type that the **PostRun** method uses to write summary data to blob storage. The **surveyAnswersSummaryCache** dictionary holds a **SurveyAnswersSummary** object for each survey.

The **PreRun** method runs before the task reads any messages from the queue and initializes a temporary cache for the new survey response data.

The **Run** method runs once for each new survey response. It uses the message from the queue to locate the new survey response, and adds the survey response to the **SurveyAnswersSummary** object for the appropriate survey by calling the **AddNewAnswer** method. The **AddNewAnswer** method updates the summary statistics in the **surveyAnswersSummaryStore** instance. The **Run** method also calls the **AppendSurveyAnswerIdToAnswersList** method to update the list of survey responses that the application uses for paging. The **Run** method leaves all the messages in the queue in case the task encounters an optimistic concurrency when it tries to save the results in the **PostRun** method.

The **PostRun** method runs after the task has invoked the **Run** method on each outstanding survey response message in the current batch. For each survey, it merges the new results with the existing summary statistics and then it saves the new values back to blob storage. The **EntitiesBlobContainer** detects any optimistic concurrency violations when it tries to save the new summary statistics and raises an exception. The **PostRun** method catches these exceptions and leaves the messages associated with current survey statistics on the queue so that they will be processed in another batch.

The worker role uses some "plumbing" code developed by Tailspin to invoke the **PreRun**, **Run**, and **PostRun** methods in the **UpdatingSurveyResultsSummaryCommand** class on a schedule. Chapter 4, "Partitioning Multi-Tenant Applications," describes this plumbing code in detail as part of the explanation about how Tailspin partitions the work in a worker role by using different tasks. The following code example shows how the Surveys application uses the "plumbing" code in the **Run** method in the worker role to run the three methods that comprise the job.

The **Run** method calls the **UpdateQueueMessage** method on the message after it has updated the list of stored survey responses to prevent a timeout from occurring that could cause the message to be reprocessed. For more information, see *"CloudQueue. UpdateMessage Method."*

```
C#
var standardQueue = this.container.Resolve
  <IAzureQueue<SurveyAnswerStoredMessage>>
  (SubscriptionKind.Standard.ToString());
var premiumQueue = this.container.Resolve
  <IAzureQueue<SurveyAnswerStoredMessage>>
  (SubscriptionKind.Premium.ToString());

BatchMultipleQueueHandler
  .For(premiumQueue, GetPremiumQueueBatchSize())
  .AndFor(standardQueue, GetStandardQueueBatchSize())
  .Every(TimeSpan.FromSeconds(
    GetSummaryUpdatePollingInterval()))
  .WithLessThanTheseBatchIterationsPerCycle(
    GetMaxBatchIterationsPerCycle())
  .Do(this.container
    .Resolve<UpdatingSurveyResultsSummaryCommand>());
```

This method first uses Unity to instantiate the **UpdatingSurveyResultsSummaryCommand** object that defines the job and the **AzureQueue** object that holds notifications of new survey responses.

The method then passes these objects as parameters to the **For**, **AndFor**, and **Do** plumbing methods of the worker role framework. The **Every** method specifies how frequently the job should run. These methods cause the plumbing code to invoke the **PreRun**, **Run**, and **PostRun** method in the **Updating-SurveyResultsSummaryCommand** class, passing a message from the queue to the **Run** method.

You should tune the frequency at which these tasks run based on your expected workloads by changing the value passed to the **Every** method.

Pessimistic and Optimistic Concurrency Handling

Tailspin uses optimistic concurrency when it saves summary statistics and survey answer lists to blob storage. The Surveys application enables developers to choose either optimistic or pessimistic concurrency when saving blobs. The following code sample from the **SurveyAnswersSummaryStore** class shows how the Surveys application uses optimistic concurrency when it saves a survey's summary statistics to blob storage.

```
C#
OptimisticConcurrencyContext context;

var id = string.Format(CultureInfo.InvariantCulture,
  "{0}-{1}", partialSurveyAnswersSummary.Tenant,
  partialSurveyAnswersSummary.SlugName);

var surveyAnswersSummaryInStore = this
  .surveyAnswersSummaryBlobContainer.Get(id, out context);

partialSurveyAnswersSummary
  .MergeWith(surveyAnswersSummaryInStore);

this.surveyAnswersSummaryBlobContainer
  .Save(context, partialSurveyAnswersSummary);
```

In this example the application uses the **Get** method to retrieve content to update from a blob. It then makes the change to the content and calls the **Save** method to try to save the new content. It passes in the **OptimisticConcurrencyContext** object that it received from the **Get** method. If the **Save** method encounters a concurrency violation, it throws an exception and does not save the new blob content.

The following code samples from the **EntitiesBlobContainer** class show how it creates a new **Optimistic-ConcurrencyContext** object in the **DoGet** method using an **ETag** object, and then uses the **Optimistic-ConcurrencyContext** object in the **DoSave** method to create a **BlobRequestOptions** object that contains the **ETag** and an access condition. The content of the **BlobRequestOptions** object enables the **UploadText** method to detect a concurrency violation; the method can then throw an exception to notify the caller of the concurrency violation.

```C#
protected override T DoGet(string objId,
          out OptimisticConcurrencyContext context)
{
  CloudBlob blob = this.Container.GetBlobReference(objId);
  blob.FetchAttributes();
  context = new OptimisticConcurrencyContext
              (blob.Properties.ETag) { ObjectId = objId };
  return new JavaScriptSerializer()
    .Deserialize<T>(blob.DownloadText());
}

protected override void DoSave(
              IConcurrencyControlContext context, T obj)
{
  ...

  if (context is OptimisticConcurrencyContext)
  {
    CloudBlob blob =
      this.Container.GetBlobReference(context.ObjectId);
    blob.Properties.ContentType = "application/json";

    var blobRequestOptions = new BlobRequestOptions()
    {
      AccessCondition =
        (context as OptimisticConcurrencyContext)
        .AccessCondition
    };

    blob.UploadText(
      new JavaScriptSerializer().Serialize(obj),
      Encoding.Default, blobRequestOptions);
  }
  else if (context is PessimisticConcurrencyContext)
  {
    ...
  }
}
```

MORE INFORMATION

All links in this book are accessible from the book's online bibliography available at: http://msdn.microsoft.com/library/jj871057.aspx.

For more information about scalability and throttling limits, see the following:

- *Windows Azure Storage Abstractions and their Scalability Targets*
- *Windows Azure SQL Database Performance and Elasticity Guide*
- *Best Practices for Performance Improvements Using Service Bus Brokered Messaging*

For more information about building large-scale applications for Windows Azure, see *"Best Practices for the Design of Large-Scale Services on Windows Azure Cloud Services"* on MSDN.

For more information about the CDN, see *"Content Delivery Network"* on MSDN.

For information about an application that uses the CDN, see the post *"EmailTheInternet.com: Sending and Receiving Email in Windows Azure"* on Steve Marx's blog.

For an episode of Cloud Cover that covers CDN, see *"Cloud Cover Episode 4 - CDN"* on Channel 9.

For a discussion of how to make your Windows Azure application scalable, see *"Real World: Designing a Scalable Partitioning Strategy for Windows Azure Table Storage."*

For more information about autoscaling in Windows Azure, see *"The Autoscaling Application Block."*

For a discussion of approaches to autoscaling in Windows Azure, see *"Real World: Dynamically Scaling a Windows Azure Application."*

For a discussion of approaches to simulating load on a Windows Azure application, see *"Real World: Simulating Load on a Windows Azure Application."*

For more information about the MapReduce algorithm, see the entry for *"MapReduce"* on Wikipedia.

6

Securing Multi-Tenant Applications

This chapter examines topics related to security in multi-tenant applications. It focuses on issues that are specific to these types of applications, such as authenticating and authorizing different sets of users that authenticate with different types of identity and through trust relationships. The chapter also discusses how you can protect individual users' data, and protect the session tokens they use when accessing your applications.

PROTECTING USERS' DATA IN MULTI-TENANT APPLICATIONS

In a multi-tenant application, tenants expect their data to be isolated from that of other tenants. A tenant will expect the application to behave as if that tenant is the sole user, and protect every tenant's private data from any unauthorized access. As tenants, they expect to own their own data and have control over who has access to it.

Authentication

Your application must determine the identity of a user and verify that the user is a tenant of the application before granting access to any private data. It is your responsibility to provide an appropriate authentication mechanism for your multi-tenant application in Windows Azure, or to enable tenants to reuse their existing authentication mechanisms.

In a multi-tenant application, tenants may also want to control and manage their own users. For example, Adatum might want four of its employees to be able to create surveys using its subscription to the Tailspin Surveys application.

In addition to defining which of their employees should have access to the application, larger tenants may also want to use their own authentication mechanism. Their employees will already have a corporate account and, rather than having to remember a new set of credentials, they would like to be able to reuse their existing corporate credentials with the new multi-tenant web hosted service. You typically implement this type of scenario by using a claims-based approach that requires you to establish trust relationships between the parties involved. For more information, see the guide *"A Guide to Claims-Based Identity and Access Control."*

Authorization

After your application has authenticated a request, it must authorize access to any resources used when it services the request. Some of the Windows Azure elements in your application may provide basic authorization services, but in most multi-tenant application scenarios you must implement the authorization yourself.

For example, there are no built-in authorization services in Windows Azure web and worker roles. If certain features implemented in your web and worker roles must be restricted to particular tenants, your application must perform the authorization based on the authenticated identity of the request.

To access Windows Azure storage services (tables, blobs, and queues), the calling code must know the storage account key for the specific storage account. Although it is unlikely that each tenant has its own storage account in a Windows Azure application, it may be the case that certain storage accounts should only be available to some tenants. Again it is your responsibility to ensure that the application code uses the correct storage account keys, and that you keep the storage account keys completely private. There is no reason for a tenant to know the storage account keys in a multi-tenant application unless the tenant is providing its own storage account.

If a tenant prefers to use a storage account in its own subscription, the tenant must provide you with the storage account keys so that the application can access the storage. It is then your responsibility to keep these keys safe on behalf of the tenant.

> *A person who gains unauthorized access to your Windows Azure account will be able to discover all of your storage account keys and access all of your data stored in Windows Azure storage. Once someone gains access to your Windows Azure subscription, there are no limits to what that person can access.*

Windows Azure Service Bus adopts a different approach and provides an authorization service to manage operations such as sending messages. Windows Azure Access Control (ACS) performs the authentication either by validating a user name and password directly, or by delegating to an external identity provider such as the tenant's Active Directory Federation Services (ADFS). For more information, see *"Service Bus Authentication and Authorization with the Access Control Service."*

Protecting Sensitive Data

As an additional safeguard in a multi-tenant application you might consider encrypting each tenant's data using a tenant specific key. This will help to ensure isolation if you can be sure that the keys used by each tenant are not revealed to anyone else.

You can use certificate based encryption in Windows Azure to strongly encrypt and decrypt data stored in Windows Azure table, blob, and queue storage; and data stored in Windows Azure SQL Database, SQL Server, and any other relational or non-relational database.

This section discusses how you decrypt data stored in Windows Azure from a worker or web role in your application. It does not address the scenario where you need to decrypt data stored in Windows Azure from a location outside of Windows Azure, such as in an on-premises application.

For sample code that illustrates how to perform encryption and decryption in a web or worker role, see the article *"Using Certificate-Based Encryption in Windows Azure Applications."* That article also describes how you should manage the private key that enables your web or worker role to decrypt your data. The important points from the article about good practices for key management are as follows:

Certificate based encryption is a standard approach to encryption that uses a key pair. You use the public key to encrypt data and the private key to decrypt data. Typically, you use X-509 certificates for these tasks. Encrypting and decrypting data in Windows Azure is easy. The challenge in Windows Azure is storing your private key securely.

- Only a small group of administrators (not developers or testers) should have access to the Windows Azure subscription that hosts the production application. These administrators are responsible for uploading a certificate that includes the private key used for decryption to the Windows Azure certificate store in the cloud service that hosts the production application.

- Under no circumstances should this certificate be available or accessible to anyone else because the private key in this certificate enables you to decrypt the data. You should have secure processes that ensure this certificate is kept secure.

- To enable a web or worker role to use the certificate for decryption you must add the certificate thumbprint to the service definition file. Typically, you add this thumbprint to the service definition file as part of an automated deployment process. Your application uses the thumbprint to locate the certificate in the Windows certificate store at runtime.

Following this approach by using the Windows Azure certificate store to store the certificate has a number of benefits:

- If the processes to manage the certificate and the way it is uploaded to Windows Azure are correctly followed, you minimize the chance that this certificate will be available to anyone who might accidentally reveal it or use it maliciously.

- There is no need for developers or testers to have access to the production certificate. They can use a different test certificate. All you need do to switch from using a test certificate to a production certificate is to update the thumbprint in the service definition file.

- If someone gains unauthorized access to the Windows Azure subscription that hosts the production application, that person cannot gain access to the private key. You cannot export a service certificate from a Windows Azure cloud service, and when Windows Azure adds the certificate to the certificate store in the role instance it marks the private key as unavailable for export.

Although this approach protects your private key, there are still some potential vulnerabilities that you must guard against:

- This approach requires you follow suitable procedures that ensure the certificate is kept secure while it is on-premises.

- Although someone who gains unauthorized access to your Windows Azure subscription cannot access the private key, they can still run code that uses the private key to decrypt any encrypted data. A malicious user could deploy their own code that reads or modifies encrypted data, or a developer could write code that accidentally reveals or changes data in a way that the application is not supposed to do.

In general, the mitigations for these risks are clear auditable procedures for managing and monitoring your Windows Azure subscription, and testing your code to ensure that it behaves in the expected way.

In other scenarios you might need to encrypt and decrypt data in on-premises applications, but store and/or share it securely in Windows Azure. For example, one organization wants to store and publish encrypted data in Windows Azure and allow selected other organizations to download and decrypt the data. This type of scenario has a different key management problem that you can address using Windows Azure Trust Services. For more information, see *"Learn More about Microsoft Codename Trust Services."*

Splitting Sensitive Data across Multiple Subscriptions

An additional technique to mitigate the risk that an attacker could discover sensitive data if your storage account keys are compromised is to split this data across two or more storage accounts. In this way, the sensitive data is not usable by the attacker unless the attacker gains access to two Windows Azure storage accounts by discovering two storage account keys.

For example, the credit card data associated with a user includes several pieces of information such as the user's name, the credit card number, the three or four digit security number, and the validity dates. Typically, to make a payment with a credit card, you must have access to all of this information. If you store the credit card numbers held by your system in one storage account, and the remaining data in a different storage account, an attacker that compromises one of the storage accounts cannot access all the information needed to use the credit cards.

However, this approach only mitigates the risk that an unauthorized person gains access to your Windows Azure storage account keys. If someone gains unauthorized access to your Windows Azure subscription, he or she can discover all of the storage account keys in that subscription. Additionally, if that Windows Azure subscription uses data from a storage account in a different Windows Azure subscription the attacker could also discover that storage account key.

Using Shared Access Signatures

In Windows Azure, knowledge of a storage account key grants full access to all of the data stored within that storage account. Therefore, if the code running a web or worker role can read a storage account key (typically from the service configuration file) it can access all the tables, blobs, and queues in that Windows Azure storage account.

You should regularly change your storage account keys, especially for storage accounts that hold sensitive data.

In many applications, allowing the web and worker roles full access to data is acceptable; but in a multi-tenant application you may want to enforce isolation by ensuring that a task or operation can only access a single tenant's data. One approach is to use a separate storage account for each tenant within the same Windows Azure subscription. However, Windows Azure limits the number of storage accounts that you can create within a single subscription, which limits the usefulness of this approach. The alternative is to use Shared Access Signatures (SAS).

A SAS is a unique and hard to guess URL that grants temporary access to a resource. For example, a SAS might grant read and write access to a specific blob for the next five minutes. To generate a SAS you must know the storage account key, but you can use the SAS without knowledge of the storage account key. Figure 1 shows how a worker role that acts as a gatekeeper can generate SAS URLs for other web and worker roles, granting them access to specific resources.

Blobs and blob containers can be configured for public read-only access so they can be read without requiring access to the storage account key.

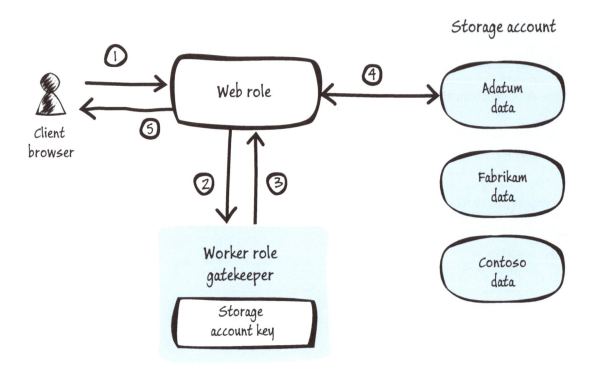

FIGURE 1
Using SAS to access data in Windows Azure storage

The following list describes the steps in Figure 1 whereby the web role gains access to the contents of the blob containing Adatum's data by using a SAS URL:

1. The client browser sends a request to view Adatum's data.

2. The web role sends a request to the gatekeeper worker role for a SAS URL that will enable read only access to Adatum's data. This data might be in table, blob, or queue storage.

3. The gatekeeper worker role uses the storage account key to generate the SAS URL and returns it to the worker role.

4. The web role uses the SAS URL when it queries for the Adatum data it needs to render the web page.

5. The web role returns the page to the browser.

There are a number of points to note about this mechanism for accessing data in Windows Azure storage:

• If the web role and the worker role are in the same cloud service they will share the same service configuration file, which means that the web role could bypass the gatekeeper and access the storage account directly. In this scenario you are relying on the code in the web role to always access storage by requesting a SAS URL.

- This approach provides for stronger isolation if the web role and the worker role are in different cloud services or different Windows Azure subscriptions. In this way you could ensure that the web role does not have access to the storage account key, and so it can access the data only by using a SAS.

- Without some additional layer of authentication and authorization, there is nothing to stop the web role asking for a SAS URL for any data.

- You can create a SAS for an individual blob or a blob container. For table storage you can create a SAS for a table, or for a set of entities stored in the table and defined using a range of partition and row keys.

- When you generate a SAS you can specify for how long it remains valid, and what types of access it supports (such as read, insert, update, and delete).

For more information see *"Creating a Shared Access Signature"* on MSDN and Chapter 5, *"Executing Background Tasks,"* in the guide *"Moving Applications to the Cloud."*

GOALS AND REQUIREMENTS

This section describes the goals and requirements for security that Tailspin has for the Surveys application.

Authentication and Authorization

The Tailspin Surveys application targets a wide range of subscribers, from individuals to large enterprises. All subscribers of the Surveys application will require authentication and authorization services to control access to their survey definitions and results, but they will want to implement these services differently.

> For more information about this scenario, see Chapter 6, *"Federated Identity with Multiple Partners,"* in the guide *"A Guide to Claims-Based Identity and Access Control."*

> You can also use a SAS to expose private blobs directly to a client. This is typically used to enhance the scalability of a solution by enabling a browser to display the content of a private blob. For more information, see Chapter 5, "Maximizing Availability, Scalability, and Elasticity."

Privacy

Tailspin wants to ensure that users' privacy is maintained. The Tailspin Surveys application should not leave any sensitive data on the client machine after a user has accessed any of the Tailspin Surveys websites. The private tenant website uses cookies to track sessions, and Tailspin wants to continue to use cookies. Therefore it has decided to encrypt the contents of these cookies in order to protect the privacy of its users.

Overview of the Solution

This section describes the approach taken by Tailspin to meet the goals and requirements that relate to security in the Surveys application.

Identity Scenarios in the Surveys Application

Tailspin has identified three different identity scenarios that the Surveys application must support:

- Organizations may want to integrate their existing identity infrastructure and be able to manage access to the Surveys application themselves in order to include Surveys as a part of the Single Sign-On (SSO) experience for their employees.

- Smaller organizations may require Tailspin to provide a complete identity system because they are not able to integrate their existing systems with Tailspin.

- Individuals and small organizations may want to re-use an existing identity, such as their Microsoft account, Open ID credentials, or an account with other social identity providers.

To support these scenarios Tailspin uses the WS-Federation protocol to implement identity federation. The following diagrams describe how the authentication and authorization process works for each of the three identity scenarios Tailspin identified.

Tailspin uses a claims-based infrastructure to provide the flexibility it needs to support its diverse subscriber base.

The three scenarios are all claims based and share the same core identity infrastructure. The only difference is the source of the original claims.

Integrating a Subscribers Own Identity Mechanism

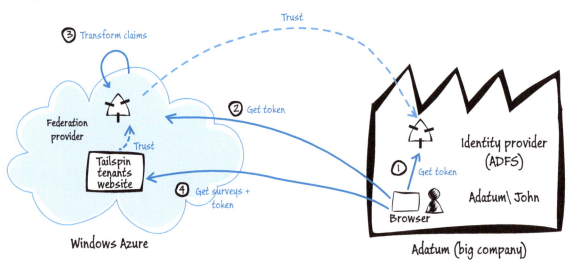

Figure 2
How users at a large enterprise subscriber access the Surveys application

In the scenario shown in Figure 2 users at Adatum, a large enterprise subscriber, authenticate with Adatum's own identity provider (step 1), in this case Active Directory Federation Services (ADFS). After successfully authenticating an Adatum user, ADFS issues a token. The client browser forwards the token to the Tailspin federation provider that trusts tokens issued by Adatum's ADFS (step 2) and, if necessary, performs a transformation on the Adatum claims in the token into claims that Tailspin Surveys recognizes (step 3) before returning a new token to the client browser. The Tailspin Surveys application trusts tokens issued by the Tailspin federation provider and uses the claims in the token to apply authorization rules (step 4).

Users at Adatum will not need to remember separate credentials to access the Surveys application, and an administrator at Adatum will be able to configure in Adatum's own ADFS the list of Adatum users that can access the Surveys application.

All of the token issuing and passing is handled automatically through a sequence of browser redirects. The user simply navigates to the Tailspin Surveys website.

Providing an Identity Mechanism for Small Organizations

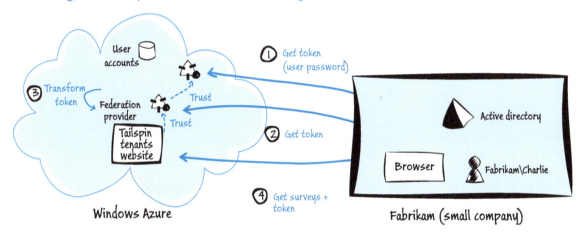

FIGURE 3
How users at a small subscriber access the Surveys application

In the scenario shown in Figure 3 users at Fabrikam, a smaller company, authenticate with the Tailspin identity provider (step 1) because their own Active Directory can't issue tokens that will be understood by the Tailspin federation provider. If the Tailspin identity provider can validate the credentials, it returns a token to the client browser that includes claims such as the user's identity and the tenant's identity. The client browser forwards the token to the Tailspin federation provider that trusts tokens issued by Tailspin identity provider (step 2) and, if necessary, performs a transformation on the Tailspin identity provider claims in the token into claims that Tailspin Surveys recognizes (step 3) before returning a new token to the client browser. The Tailspin Surveys application trusts tokens issued by the Tailspin federation provider and uses the claims in the token to apply authorization rules (step 4).

Other than the choice of identity provider, this approach is the same as that used for Adatum. The downside of this approach for Fabrikam users is that they must memorize separate credentials just to access the Surveys application. Fabrikam users will be prompted for their credentials when they navigate to the Tailspin Surveys application. Tailspin must also provide a way to manage the user accounts that the Tailspin identity provider uses.

Tailspin plans to implement this scenario by using an ASP.NET membership provider to manage the user accounts, and use a security token service (STS) that integrates with the membership provider.

> *For guidance on how to implement this scenario take a look at the **thinktecture IdentityServer** project on CodePlex.*

Integrating with Social Identity Providers

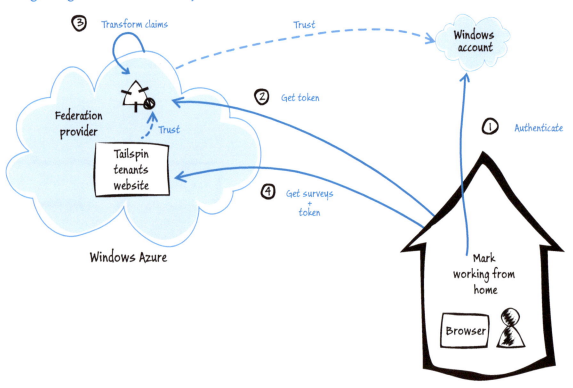

FIGURE 4
How an individual subscriber accesses the Surveys application

For individual users the process is again very similar. In the scenario shown in Figure 4 the Tailspin federation provider is configured to trust tokens issued by a third-party identity provider, such as an identity provider that authenticates a Microsoft account or OpenID credentials. Tailspin plans to use Windows Azure Access Control to implement this scenario.

When an individual user authenticates with his or her chosen identity provider (step 1), the identity provider returns a token to the client browser that includes claims such as the user's identity. The client browser forwards the token to the Tailspin federation provider that trusts tokens issued by the third-party provider (step 2) and, if necessary, performs a transformation on the claims in the token into claims that Tailspin Surveys recognizes (step 3) before returning a new token to the client browser. The Tailspin Surveys application trusts tokens issued by the Tailspin federation provider and uses the claims in the token to apply authorization rules (step 4). When the user tries to access their surveys, the application will redirect them to their external identity provider for authentication.

> *For additional guidance on how to implement this scenario, see the chapter "Federated Identity with Multiple Partners and Windows Azure Access Control Service" in the guide "A Guide to Claims-Based Identity and Access Control."*

Windows Azure Access Control Service and Windows Azure Active Directory

Although the Tailspin Surveys sample solution uses Windows Identity Foundation (WIF) to implement a WS-Federation compliant federation provider (see the TailSpin.SimulatedIssuer project in the solution), a production deployment would use a real federation provider such as Active Directory Federation Services (ADFS), Windows Azure Access Control, or Windows Azure Active Directory.

Windows Azure Access Control is one element of Windows Azure Active Directory. It enables you to move authentication and authorization logic out of your code and into a separate cloud-based service. Access Control can integrate with other standards based identity providers, and can implement a claims transformation process using declarative rules that convert the claims issued by the tenant's issuer, third-party issuer, or Tailspin's own issuer into claims understood by the Tailspin Surveys application. Access Control can also perform the protocol conversion required to support many third-party issuers.

Windows Azure Active Directory includes the Windows Azure Authentication Library that allows developers to focus on business logic in their applications, ignore most protocol details, and easily secure resources without being an expert on security. Windows Azure Active Directory also includes a REST API that enables programmatic access to Access Control and the Authentication Library.

> *For more information see "Windows Azure Active Directory." For more information about using Access Control see the related patterns & practices "Claims Based Identity & Access Control Guide."*

Configuring Identity Federation for Tenants

When a new tenant subscribes to the Tailspin Surveys service, it has the option to use its own identity provider instead of Tailspin's to authenticate its users when accessing the private tenant web site. In the sample application code, the Join screen is currently mocked out to illustrate the information that a tenant would need to provide in order to establish the federated identity environment described earlier in this chapter.

> In a real application this screen would enable a tenant to select between the three identity scenarios supported by Tailspin Surveys. For tenants who chose to use the Tailspin identity provider you would have a registered members database and each tenant would be allowed to add and remove members authorized to use the subscription.

However, the sample application does allow a Tailspin administrator to add a new federated identity provider on behalf of a tenant on the Manage screen. Tailspin Surveys then saves the configuration data in Windows Azure blob storage as part of the tenant configuration information. The following table describes the information used to configure identity federation for a tenant.

Value	Description	Example
Identifier	Tailspin's identity provider uses this value to recognize claims sent from a trusted identity provider.	http://adatum/trust
Sign-in URL	The address of the tenant's trusted identity provider.	https://localhost/Adatum.SimulatedIssuer.v2/
Thumb-print	The thumbprint of the certificate used by the tenant's identity provider to sign the claims it sends to Tailspin.	f260042d59e14817984c6183fbc6bfc71baf5462
Admin Claim Type	The claim type that the tenant uses to identify users with administrative privileges in their Tailspin Surveys subscription. This is used to map the tenant's claim type to the Tailspin **Role** claim type.	http://schemas.xmlsoap.org/claims/group
Admin Claim Value	The value of the administrator claim type that has administrative privileges in the tenant's Tailspin Surveys subscription. This value is mapped to the **SurveyAdministrator** role in Tailspin Surveys.	Marketing Managers

Encrypting Session Tokens in a Windows Azure Application

The Tailspin Surveys tenant web site uses sessions to maintain the list of questions that a tenant adds when designing a new survey. The website uses a cookie to track requests that belong to the current user's session. Part of Tailspin's security requirements is that the application should encrypt cookies so that there is no usable information left on the client machine.

Tailspin plans to use at least two instances of the web role that hosts the tenant website in order to make the site more available. Therefore, the encryption mechanism that the application uses to encrypt the cookies must be web farm friendly. A cookie that one role instance creates and encrypts must be readable by all other instances.

By default, when you use the Windows Identity Foundation (WIF) framework to manage your identity infrastructure, it encrypts the contents of the cookies that it sends to the client by using the Windows Data Protection API (DPAPI). Using DPAPI for cookie encryption is not a workable solution for an application that has multiple role instances because each role instance will use a different encryption key, and the Windows Azure load balancer could route a request to any instance. You must use an encryption mechanism such as RSA, which uses a key that two or more instances can share.

INSIDE THE IMPLEMENTATION

Now is a good time to walk through some of the code in the Tailspin Surveys application in more detail. As you go through this section, you may want to download the Visual Studio solution for the Tailspin Surveys application from *http://wag.codeplex.com/*.

Using Windows Identity Foundation

Figure 5 will help you to keep track of how the WIF authentication process works as you look at the detailed description and code samples later in this chapter.

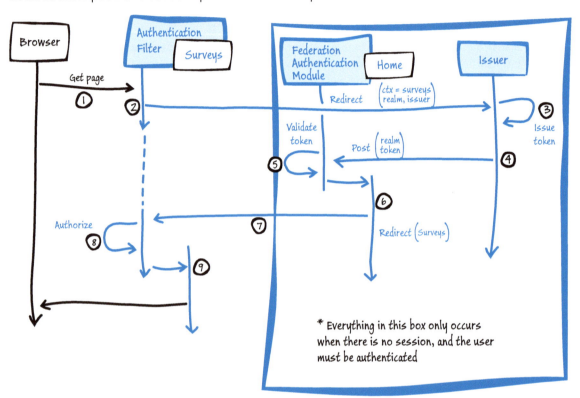

FIGURE 5
Federation with multiple partners sequence diagram

For clarity, Figure 5 shows the "logical" sequence, not the "physical" sequence. Wherever the diagram has an arrow with a **Redirect** label, this actually sends a **redirect** response back to the browser, and the browser then sends a request to wherever the **redirect** message specifies.

The following describes the steps illustrated in Figure 5:

1. The process starts when an unauthenticated user sends a request for a protected resource; for example the adatum/surveys page. This invokes a method in the **SurveysController** class.

The sequence shown in Figure 5 applies to all three authentication and identity scenarios described earlier in this chapter. In the context of Figure 5, the Issuer is the Tailspin federation provider, so step 3 includes redirecting to another issuer to handle the authentication.

2. The **AuthenticateAndAuthorizeTenant** attribute that extends the **AuthenticateAndAuthorize-Role** attribute and implements the MVC **IAuthorizationFilter** interface is applied to this controller class. Because the user has not yet been authenticated, this will redirect the user to the Tailspin federation provider at https://localhost/TailSpin.SimulatedIssuer with the following query string parameter values:

wa. Wsignin1.0

wtrealm. https://tailspin.com

wctx. https://127.0.0.1:444/survey/adatum

whr. http://adatum/trust

wreply. https://127.0.0.1:444/federationresult

The following code example shows the **BuildSignInMessage** method in the **AuthenticateAndAuthorizeTenantAttribute** class that builds the query string.

```csharp
protected override WSFederationMessage
  BuildSignInMessage(AuthorizationContext context,
                     Uri replyUrl)
{
  var tenant =
    (context.Controller as TenantController).Tenant;

  var fam = FederatedAuthentication
            .WSFederationAuthenticationModule;
  var signIn = new SignInRequestMessage
                (new Uri(fam.Issuer), fam.Realm)
  {
    Context = AuthenticateAndAuthorizeRoleAttribute
      .GetReturnUrl(context.RequestContext,
                    RequestAppendAttribute.RawUrl,
                    null).ToString(),
    HomeRealm = SubscriptionKind.Premium
      .Equals(tenant.SubscriptionKind)
      ? tenant.IssuerIdentifier
      ?? Tailspin.Federation.HomeRealm + "/"
      + (context.Controller
        as TenantController).Tenant.Name
      : Tailspin.Federation.HomeRealm + "/"
      + (context.Controller
        as TenantController).Tenant.Name,
    Reply = replyUrl.ToString()
  };

  return signIn;
}
```

3. The Issuer, in this case the Tailspin simulated issuer, authenticates the user and generates a token with the requested claims. In the Tailspin scenario, the Tailspin federation provider uses the value of the **whr** parameter to delegate the authentication to another issuer; in this example, the Adatum issuer. If necessary, the Tailspin federation issuer can transform the claims it receives from the issuer into claims that the Tailspin Surveys application understands. The following code from the **FederationSecurityTokenService** class shows how the Tailspin simulated issuer transforms the **Group** claims in the token from the Adatum issuer.

```csharp
C#
protected override IClaimsIdentity
  GetOutputClaimsIdentity(IClaimsPrincipal principal,
                    RequestSecurityToken request,
                    Scope scope)
{
  ...

  var input = principal.Identity as ClaimsIdentity;

  var tenant = this.tenantStore.GetTenant
    (input.Claims.First().Issuer);

  ...

  var output = new ClaimsIdentity();

  CopyClaims(input,
    new[] { WSIdentityConstants.ClaimTypes.Name },
    output);
  TransformClaims(input, tenant.ClaimType,
    tenant.ClaimValue, ClaimTypes.Role,
    Tailspin.Roles.SurveyAdministrator, output);
  output.Claims.Add(
    new Claim(Tailspin.ClaimTypes.Tenant,
              tenant.Name));

  return output;
}
```

*This example shows how the claim the tenant identified as granting access to the subscription is mapped to the Tailspin **Role** claim with a value of **SurveyAdministrator**.*

4. The Tailspin federation provider then posts the token and the value of the *wctx* parameter (https://127.0.0.1:444/survey/adatum) back to the address in the *wreply* parameter (https://127.0.0.1:444/federationresult). This address is another MVC controller, which does not have the **AuthenticateAndAuthorizeTenantAttribute** attribute applied. The following code example shows the **FederationResult** method in the **ClaimsAuthenticationController** controller.

```C#
[RequireHttps]
public class ClaimsAuthenticationController
  : Controller
{
  [ValidateInput(false)]
  [HttpPost]
  public ActionResult FederationResult()
  {
    var fam = FederatedAuthentication
            .WSFederationAuthenticationModule;
    if (fam.CanReadSignInResponse(
      System.Web.HttpContext.Current.Request, true))
    {
      string returnUrl = GetReturnUrlFromCtx();
      return this.Redirect(returnUrl);
    }
    return this.RedirectToAction(
            "Index", "OnBoarding");
  }
```

5. The WS Federation Authentication Module validates the token by calling the **CanReadSignIn-Response** method.

6. The **ClaimsAuthenticationController** controller retrieves the value of the original *wctx* parameter and issues a redirect to that address.

7. This time, when the request for the adatum/surveys page goes through the **AuthenticateAnd-AuthorizeTenantAttribute** filter, the user has been authenticated. The following code example from the **AuthenticateAndAuthorizeRoleAttribute** class shows how the filter checks whether the user is authenticated.

```C#
public void OnAuthorization(
            AuthorizationContext filterContext)
{
  ...

  if (!filterContext.HttpContext.User
      .Identity.IsAuthenticated)
  {
    AuthenticateUser(filterContext);
  }
  else
  {
    this.AuthorizeUser(filterContext);
  }
  ...
}
```

8. The **AuthenticateAndAuthorizeTenantAttribute** filter then applies any authorization rules. In the Tailspin Surveys application the **AuthorizeUser** method verifies that the user is a member of one of the roles listed where the **AuthenticateAndAuthorize-Tenant** attribute decorates the MVC controller, as shown in the following code example.

C#
```
[AuthenticateAndAuthorizeTenant(
                Roles = "Survey Administrator")]
[RequireHttps]
public class SurveysController : TenantController
{
  ...
}
```

9. Finally, the controller method executes.

Protecting Session Tokens in Windows Azure

The following code example shows how the Surveys application configures the session security token handler to use RSA encryption instead of the default DPAPI encryption. This enables Tailspin to deploy multiple instances of the web role that can use the shared key.

Before Tailspin deploys the Surveys application to Windows Azure it must upload the shared encryption key to the Windows Azure certificate store. Windows Azure stores the key as part of the cloud service definition, and it is accessible to all the roles and role instances deployed to the cloud service.

An ASP.NET web application running in an on-premises web farm would also need to use shared key encryption instead of DPAPI.

> *You can create an X-509 certificate that is suitable for use with the RSA encryption algorithm by using the makecert tool. For more information, see "How to Create a Certificate for a Role." For more information about uploading a key to the Windows Azure certificate store, see "How to Add a New Certificate to the Certificate Store."*

The following code from the Global.asax file in the Tailspin.Web project shows how the application loads the certificate it will use for encrypting and decrypting the session cookie from the certificate store in the cloud service. The application identifies the certificate from the thumbprint in the **serviceCertificate** element in the Web.config file.

```csharp
C#
private void OnServiceConfigurationCreated(object sender,
    ServiceConfigurationCreatedEventArgs e)
{
  var sessionTransforms =
      new List<CookieTransform>(
          new CookieTransform[]
          {
            new DeflateCookieTransform(),
            new RsaEncryptionCookieTransform(
              e.ServiceConfiguration.ServiceCertificate),
            new RsaSignatureCookieTransform(
              e.ServiceConfiguration.ServiceCertificate)
          });
  var sessionHandler = new SessionSecurityTokenHandler(
                      sessionTransforms.AsReadOnly());
  e.ServiceConfiguration.SecurityTokenHandlers
                    .AddOrReplace(sessionHandler);
}
```

The **Application_Start** method in the Global.asax.cs file hooks up this event handler to the **FederatedAuthentication** module.

For more information about the DPAPI, see "Windows Data Protection" on MSDN.

MORE INFORMATION

All links in this book are accessible from the book's online bibliography available at: *http://msdn.microsoft.com/library/jj871057.aspx.*

For more information about the claims-based authentication and authorization model used in the Surveys application see Chapter 6, *"Federated Identity with Multiple Partners,"* of the guide *"A Guide to Claims-Based Identity and Access Control."*

For a walkthrough of how to secure an ASP.NET site on Windows Azure with WIF, see *"Exercise 1: Enabling Federated Authentication for ASP.NET applications in Windows Azure"* on Channel 9.

For more information about using forms authentication with your Windows Azure application, see *"Real World: ASP.NET Forms-Based Authentication Models for Windows Azure."*

7 Managing and Monitoring Multi-Tenant Applications

This chapter discusses two main areas of concern when you build and deploy multi-tenant applications. The first is related to application lifecycle management (ALM) and covers topics such as testing, deployment, management, and monitoring. The second is related specifically to independent software vendors (ISVs) that are building multi-tenant applications, and discusses topics such as how to enable onboarding, customization, and billing for tenants and customers that use the application.

ALM CONSIDERATIONS FOR MULTI-TENANT APPLICATIONS

All applications require a consistent policy for their lifecycle to ensure that development, testing, deployment, and management are integrated into a reliable and repeatable process. This helps to ensure that applications work as expected, provide all the required features, and operate efficiently and reliably.

However, there are some additional considerations for multi-tenant applications. For example, you may need to implement more granular management, monitoring, and update procedures based around the separation for each tenant. This may include backing up individual tenant's data separately to minimize security concerns, or being able to update specific instances of the application that are reserved for some tenants.

Goals and Requirements

Tailspin's ALM goals and requirements for the Surveys application encompass those that are applicable to most multi-tenant applications.

Managing the application lifecycle for multi-tenant applications is usually more complex than that for other types of application because some tasks must be carried out on a per-tenant basis.

In terms of testability, there are two main areas that must be addressed: unit testing of application components both during and after development, and functional testing of the application or sections of it in a realistic runtime environment. Tailspin must address both of these areas by designing the application to maximize testability. The developers at Tailspin want to implement the application in a way that allows them to use mock objects to simplify unit tests. They also want to support testing for sections of the application, such as the background tasks carried out by worker roles, as well as being able to quickly and easily deploy the application with a test configuration to a staging platform for functional and acceptance testing.

Tailspin wants to be able to perform as much testing as possible using the local compute and storage emulators. Of course, Tailspin will still do a complete test pass in Windows Azure before the final deployment, but during development it is more convenient to test locally. This is one of the factors that Tailspin will consider when it makes technology choices. For example, if Tailspin uses Windows Azure Caching instead of Windows Azure Shared Caching, it can test the caching behavior of the application locally. However, if Tailspin chooses to use SQL Database federations then it cannot test that part of the application locally.

In addition to the unit tests and functional testing, Tailspin wants to verify that the Surveys application will scale to meet higher levels of demand, and to determine what scale units it should use. For example, how many worker role instances and message queues should there be for every web role instance when Tailspin scales out the application. There is little point in performing stress testing using the local compute and storage emulators; you must deploy the application to the cloud in order to determine how it performs under realistic load conditions.

When testing is complete, Tailspin's administrators want to be able to deploy the application in a way that minimizes the chance of error. This means that they must implement reliable and repeatable processes that apply the correct configuration settings and carry out the deployment automatically. They also want to be able to update the application while it is running, and roll back changes if something goes wrong.

After the application has been successfully deployed, the administrators and operators at Tailspin must be able to manage the application while it is running. This includes tasks such as backing up data, adjusting configuration settings, managing the number of instances of roles, handling requests for customization, and more. In a multi-tenant application, administrators and operators may also be responsible for all or part of the onboarding process for trials and new subscribers.

You should check the latest documentation for both the compute and storage emulators to identify any differences in their behavior from the Windows Azure services.

Finally, administrators and operators must be able to monitor the application to ensure that it is operating correctly, meeting its SLAs, and fulfilling business requirements. For a multi-tenant application, administrators will also want to be able to monitor both the operation and the runtime costs on a per-tenant basis. Where the application supports different levels of SLA or feature set for different types of tenants, monitoring requirements can become considerably more complicated. For example, heavy loading on specific instances will often require the deployment of additional instances to ensure that SLAs are met. Tasks such as this will typically require some kind of automation that combines the results of monitoring with the appropriate management actions.

Procedures for managing and monitoring multi-tenant applications and the data they use must take into account the requirements of individual tenants to maintain security and to meet SLAs.

Overview of the Solution

This section describes the options Tailspin considered for testing, deploying, managing, and monitoring the Surveys application, and identifies their chosen solutions.

Testing Strategies

One of the advantages with a multi-tenant application compared to multiple different application implementations is that there is only a single code base. Every tenant runs the same core application code, and so there is only one application and one set of components to test.

However, because most multi-tenant applications support user customization through configuration, the test processes must exercise all of the configurable options. Unit tests, functional tests, and acceptance testing must include passes that exercise every possible combination of configurable options to ensure that all work correctly and do not conflict with each other.

If tenants can upload their own components or resources (such as style sheets and scripts) testing should encompass this to as wide an extent as possible. While it will not be possible to test every component or resource, the tests should ensure that these cannot affect execution of the core application code, or raise security issues by exposing data or functionality that should not be available.

> When testing features that allow users to upload code, scripts, or style sheets your test process should resemble that of a malicious user by uploading items that intentionally attempt to access data and interfere with execution. While you cannot expect to cover every eventuality, this will help to expose possible areas where specific security measures must be applied.

In terms of test environment requirements, multi-tenant applications are generally no different from any other application. Developers and testers run unit tests on local computers and on the build server to validate individual parts of the application. This includes using the local Windows Azure compute and storage emulators on the development and test computers. The one area that may require additional test environment capacity is to provide separate resources, such as databases, to represent multiple tenant resources during testing.

Functional tests are run in a local test environment that mirrors the Windows Azure runtime environment as closely as possible; or in a staging environment on Windows Azure. Typically this will use a different subscription from the live environment to ensure that only administrators and responsible personnel have access to the keys needed for deployment and access to live services and resources such as databases.

Acceptance tests occur after final deployment, and the Windows Azure capability for rolling back changes through a virtual IP swap allows rapid reversion to a previous version should a failure occur. Acceptance testing must include testing the application from the user's perspective, including applying any customizations that are supported.

Other types of tests such as performance, throughput, and stress testing are carried out as part of the functional tests, and throughout final testing, to ensure that the application can meet its SLAs.

Stress testing should include verifying that any autoscaling rules add or remove sufficient resources to support the level of demand and control your costs.

Designing to Support Unit Testing

The Surveys application uses Windows Azure table and blob storage, and the developers at Tailspin were concerned about how this would affect their unit testing strategy. From a testing perspective, a unit test should focus on the behavior of a specific class and not on the interaction of that class with other components in the application. From the perspective of Windows Azure, any test that depends on Windows Azure storage requires complex setup and tear-down logic to make sure that the correct data is available for the test to run.

For both of these reasons, the developers at Tailspin designed the data access functionality in the Surveys application with testability in mind, and specifically to make it possible to run unit tests against their data store classes without a dependency on Windows Azure storage.

The Surveys application uses the Unity Application Block to decouple its components and facilitate testing.

The solution adopted by the developers at Tailspin was to wrap the Windows Azure storage components in such a way as to facilitate replacing them with mock objects during unit tests, and use the Unity Application Block to instantiate them. A unit test should be able to instantiate a suitable mock storage component, use it for the duration of the test, and then discard it. Any integration tests can continue to use the original data access components to test the functionality of the application.

> Unity is a lightweight, extensible dependency injection container that supports interception, constructor injection, property injection, and method call injection. You can use Unity in a variety of ways to help decouple the components of your applications, to maximize coherence in components, and to simplify design, implementation, testing, and administration of these applications. You can learn more about Unity and download the application block from *"Unity Container."*

Tailspin also wanted to be able to separately unit test the background tasks implemented as Windows Azure worker roles. Tailspin's developers created a generic worker role framework for the worker roles that makes it easy to add and update background task routines, and also supports unit testing.

The worker role framework Tailspin implemented allows individual jobs to override the **PreRun**, **Run**, and **PostRun** methods to set up, run, and tear down each job. The support in this framework for operators such as **For**, **Do**, and **Every** to execute tasks in a worker also makes it easy to write unit tests for jobs that will be processed by a worker role. Chapter 4, "Partitioning Multi-Tenant Applications," describes the implementation of these operators, and the section "Testing Worker Roles" later in this chapter illustrates how they facilitate designing unit tests.

Stress Testing and Performance Tuning

Tailspin performed stress testing on the Surveys application running in the cloud in order to uncover any bottlenecks that limit the application's scalability, and to understand how to scale out the application. Bottlenecks might include limits on the throughput that can be achieved with Windows Azure storage, limits on the amount of processing that the application can perform with given the available CPU and memory resources and the algorithms used by the application, and limits on the number of web requests that the web roles can handle.

After a stress test identified a bottleneck, the team at Tailspin evaluated the available options for removing the bottleneck, made a change to the application, and then re-ran the stress test to verify that the change had the expected effect on the application.

To perform the stress testing, Tailspin used Visual Studio Load Test running in Windows Azure to simulate different volumes of survey response submissions to the public Surveys web site. For more information about how to run load tests in Windows Azure roles that exercise another Windows Azure application, see *Using Visual Studio Load Tests in Windows Azure Roles* on MSDN.

Application Deployment and Update Strategies

Multi-tenant application deployment and updating follows the same process as other types of applications. There should be only one core code package for the application because all customization for individual tenants should be accomplished through just configuration settings for each tenant. These configuration settings should ideally be stored in a separate location, such as a database or Windows Azure storage, and not in the service configuration files. This removes the requirement to upload different versions of the application.

Where tenants have the ability to upload additional resources, such as style sheets and logos, these resources should be stored outside of the application. Therefore, deploying a new application or updating an existing one will not impact individual tenant's resources. The updated application will continue to read configuration settings from the central configuration store and access the tenant's resources from the central location where they are stored.

To provide a reliable and repeatable deployment and update experience, Tailspin uses scripts that run as part of the build process when deploying to the test environment, and separate scripts accessible only to administrators that are executed to deploy or update the application in the live runtime environment. This prevents the possibility of errors that could occur when using the Windows Azure Management Portal.

The scripts modify the settings in the Web.config file that cannot be stored in the service configuration files, such as the settings for Windows Identity Foundation (WIF) authentication. The scripts also accept a parameter that defines which of the service configuration files will be uploaded to Windows Azure during deployment or update. There are separate service configuration files in the source code project for use when deploying to the local and cloud test environments.

> *See Chapter 3, "Moving to Windows Azure Cloud Services," of the guide "Moving Applications to the Cloud" for more details of this approach to deploying to a local testing environment, a cloud based staging area, and the live environment.*

Application Management Strategies

Windows Azure incorporates several features that are useful for managing applications after deployment. These include the Windows Azure Management Portal, the Windows Azure Management API, the Windows Azure PowerShell cmdlets, and many Microsoft and third party tools and services.

Although you can modify settings in your Windows Azure application's service configuration file (.cscfg) on the fly, you cannot add new settings without redeploying the role. In a multi-tenant application, you typically require new settings for each tenant. Therefore, Tailspin chose to store tenant configuration data in blob storage and read it from there whenever the application needs it.

Administrators can use the Windows Azure Management Portal to modify configuration settings in the service configuration file while the application is deployed; and to stop and start roles, change the number of instances, and see basic runtime information about roles. All of these tasks can also be accomplished by using the Windows Azure Management API, and administrators can use a series of Power-Shell cmdlets that provide a wide range of methods for interacting with the API.

> *You can obtain the Windows Azure PowerShell cmdlets from the Windows Azure **Download** Page.*

Tailspin uses the Windows Azure PowerShell cmdlets to interact with the Windows Azure Management API for almost all management tasks. This provides a reliable and repeatable process for common tasks that administrators must carry out. However, for some tasks administrators will use the Windows Azure Management Portal— particularly for tasks that are not carried out very often.

Reliability and Availability

One of the major concerns for administrators and operators who manage the application is to ensure that it is available at all times and is meeting its SLAs. It is very difficult to estimate the workload for a multi-tenant application because the tenants are unlikely to provide detailed estimates of usage, and the peaks can occur at various times if users are located in many different time zones.

Tailspin realized that one of the core factors for meeting SLAs would be to ensure sufficient instances of the application are running at all times, while minimizing costs by removing instances when not required. To achieve this Tailspin will incorporate the Enterprise Library Autoscaling Application Block into the application to automatically add or remove role instances based on changes in average load and usage of the application.

To achieve a base level of reliability Tailspin always ensures that a minimum of two instances of each role are deployed so that failures, or reorganization of resources within the Windows Azure datacenter, will not prevent users from accessing the application.

> *See Chapter 5, "Maximizing Availability, Scalability, and Elasticity," for more information about how Tailspin plans to use the Enterprise Library Autoscaling Application Block in Tailspin Surveys.*

Tailspin plans to keep detailed usage information that it can analyze for trends. If Tailspin can identify times of the day, week, month, or year when usage is regularly higher or lower it can preemptively add or remove resources. The Autoscaling Application Block enables Tailspin to perform this type of autoscaling in addition to reactive scaling based on average load or usage.

Backup and Restore for Data

The most significant changes to administrative and management tasks for a multi-tenant application when compared to a standard business application are concerned with the processes used to back up and restore data. Each tenant's data will typically be isolated from all others through the use of a separate database, table, partition, blob container, or storage account. It is vital that this isolation is not compromised during the backup and restore process.

If the data is held in separate databases, the backup procedures can simply address each database in turn, and store the backup in separate files or blobs. However, these files or blobs must also be securely stored so that only responsible staff and the appropriate tenant can access them. A tenant must obviously not be able to access a backup that contains other tenants' data.

Subscribers who export their survey data to SQL Database can use the SQL Database Import/Export Service to back up their data. This service enables you to export your data to blob storage and then optionally download it to an on-premises location for safekeeping.

If tenants' data is stored in separate Windows Azure subscriptions you must consider whether your responsibility includes backup and restore processes. One of the reasons that tenants may want to use their own subscription and account for their data storage is to maximize security of the data or to meet regulatory limitations (such as on the location or storage of the data). In most cases, the tenant should be responsible for backing up and restoring this data.

If all of the tenants' data is held in shared storage, such as in a shared database or in a single Windows Azure storage account, specific care must be taken when designing the backup and restore processes. While it is possible to create a single backup of the database or storage account, doing so means that administrators must exercise extra care in storing the backup and when restoring some or all tenants' data. One solution may be to offer the capability in the application for tenants to create a backup containing just their own data on demand, and allow them to store it in a location of their choosing.

In the Surveys application, data for each tenant is stored in Windows Azure tables. Tailspin will implement a mechanism that allows tenants to back up the data for each tenant separately, and store each tenant's backup in a separate Windows Azure storage blob to maintain isolation. For examples illustrating how to backup Windows Azure table storage, see *"Table Storage Backup & Restore for Windows Azure"* on CodePlex, and the blog post *"Protecting Your Tables Against Application Errors."*

There are several other management-related concerns particularly applicable to ISVs and software vendors, rather than to the development of in-house applications and services. These include how the application supports on-boarding and configuration for subscribers, per tenant customization, and financial goals. These topics are discussed in more detail in the section "ISV Considerations for Multi-Tenant Applications" later in this chapter.

Application Monitoring Strategies

Windows Azure incorporates several features that are useful for monitoring applications. These include the Windows Azure Management Portal, the Windows Azure Management API, the Windows Azure diagnostics mechanism, and many tools and services available from Microsoft and from third parties. For example, Microsoft System Center can be used to monitor a Windows Azure application and raise alerts when significant events occur.

Developers should make use of the Windows Azure diagnostics mechanism to generate error and trace messages within the application code. In addition, administrators can configure Windows Azure diagnostics to record operating system events and logs, and other useful information. All of the monitoring information collected by the diagnostics mechanism can be accessed using a range of tools to view the Windows Azure tables and blobs where this data is stored, or by using scripts to download the data for further analysis.

Where tenants can upload resources to customize the application, administrators can take advantage of Windows Azure endpoint protection to guard against the occurrence of malicious code such as viruses and Trojans finding their way onto the server. You can install endpoint protection into each web and worker role instance in your application and then configure Windows Azure diagnostics to read error and warning messages from the Microsoft Antimalware source in the system event log.

> See "Microsoft Endpoint Protection for Windows Azure" on the Microsoft download site for more information. On this page you can also download the document "Monitoring Microsoft Endpoint Protection for Windows Azure," which describes how to collect diagnostic data from Windows Azure endpoint protection.

In a multi-tenant application you must take special care when dealing with log files because diagnostic data can include tenant specific data. If you allow tenants to see log files, perhaps for troubleshooting purposes, you must be sure that diagnostic data is not shared accidentally with the wrong tenant. You must either keep separate logs for each tenant, or be sure that you can filter logs by tenant before sharing them.

Tailspin implements code in the application that writes events to Windows Azure diagnostics by using a custom helper class and the Windows Azure Diagnostics listener. This includes a range of events and covers common error situations within the application. A configuration setting in the service configuration file controls the level of event logging, allowing administrators to turn on extended logging when debugging the application and turn it off during ordinary runtime scenarios.

Inside the Implementation

Now is a good time to walk through some of the code in the Tailspin Surveys application in more detail. As you go through this section, you may want to download the Visual Studio solution for the Tailspin Surveys application from *http://wag.codeplex.com/*.

Unit Testing

Tailspin designed many classes of the Surveys application to support unit testing by taking advantage of the dependency injection design pattern. This allows mock objects to be used when testing individual classes without requiring the complex setup and teardown processes often needed to use the real objects.

For example, this section describes how the design of the Surveys application supports unit testing of the **SurveyStore** class that provides access to Windows Azure table storage. This description focuses on tests for one specific class, but the application uses the same approach with other store classes.

The following code example shows the **IAzureTable** interface and the **AzureTable** class that are at the heart of the implementation.

```C#
public interface IAzureTable<T> :
  IAzureObjectWithRetryPolicyFactory
  where T : TableServiceEntity
{
  IQueryable<T> Query { get; }
  CloudStorageAccount Account { get; }

  void EnsureExist();
  void Add(T obj);
  void Add(IEnumerable<T> objs);
  void AddOrUpdate(T obj);
  void AddOrUpdate(IEnumerable<T> objs);
  void Delete(T obj);
  void Delete(IEnumerable<T> objs);
}

public class AzureTable<T> : AzureStorageWithRetryPolicy,
  IAzureTable<T> where T : TableServiceEntity
{
  private readonly string tableName;
  private readonly CloudStorageAccount account;

  ...
```

```csharp
public IQueryable<T> Query
{
  get
  {
    TableServiceContext context = this.CreateContext();
    return context.CreateQuery<T>(this.tableName)
      .AsTableServiceQuery();
  }
}

...

public void Add(T obj)
{
  this.Add(new[] { obj });
}

public void Add(IEnumerable<T> objs)
{
  TableServiceContext context = this.CreateContext();

  foreach (var obj in objs)
  {
    context.AddObject(this.tableName, obj);
  }

  var saveChangesOptions = SaveChangesOptions.None;
  if (objs.Distinct(
    new PartitionKeyComparer()).Count() == 1)
  {
    saveChangesOptions = SaveChangesOptions.Batch;
  }

  this.StorageRetryPolicy.ExecuteAction(()
    => context.SaveChanges(saveChangesOptions));
}

...

private TableServiceContext CreateContext()
{
  return new TableServiceContext(
    this.account.TableEndpoint.ToString(),
    this.account.Credentials)
  {
    // Retry policy is handled by TFHAB
    RetryPolicy = RetryPolicies.NoRetry()
  };
}
```

```
  private class PartitionKeyComparer :
    IEqualityComparer<TableServiceEntity>
  {
    public bool Equals(TableServiceEntity x,
      TableServiceEntity y)
    {
      return string.Compare(x.PartitionKey,
        y.PartitionKey, true,
        System.Globalization.CultureInfo.InvariantCulture)
        == 0;
    }

    public int GetHashCode(TableServiceEntity obj)
    {
      return obj.PartitionKey.GetHashCode();
    }
  }
}
```

*The **Add** method that takes an **IEnumerable** parameter should check the number of items
in the batch and the size of the payload before calling the SaveChanges method with the
SaveChangesOptions.Batch option. For more information about batches and Windows Azure
table storage, see "Performing Entity Group Transactions" on MSDN.*

The generic interface and class have a type parameter **T** that derives from the Windows Azure **Table-
ServiceEntity** type you use to create your own table types. For example, in the **Surveys** application
the **SurveyRow** and **QuestionRow** types derive from the **TableServiceEntity** class. The **IAzureTable**
interface defines several operations: the **Query** method returns an **IQueryable** collection of the type
T, and the **Add**, **AddOrUpdate**, and **Delete** methods each take a parameter of type **T**. In the **Azure-
Table** class the **Query** method returns a **TableServiceQuery** object, the **Add** and **AddOrUpdate**
methods save the object to table storage, and the **Delete** method deletes the object from table
storage.

To create a mock object for unit testing, you must instantiate an object that implements the interface
type **IAzureTable**. The following code example from the **SurveyStore** class shows the constructor.
Because the constructor takes parameters of type **IAzureTable**, you can pass in either real or mock
objects that implement this interface.

```C#
public SurveyStore(IAzureTable<SurveyRow> surveyTable,
                   IAzureTable<QuestionRow> questionTable)
{
  this.surveyTable = surveyTable;
  this.questionTable = questionTable;
}
```

This parameterized constructor is invoked in two different scenarios. The Surveys application invokes it indirectly when the application uses the **SurveysController** MVC class. The application uses the Unity dependency injection framework to instantiate MVC controllers. The Surveys application replaces the standard MVC controller factory with the **UnityControllerFactory** class in the **OnStart** method in both web roles, so when the application requires a new MVC controller instance Unity is responsible for instantiating that controller. The following code example shows part of the **ContainerBootstrapper** class from the TailSpin.Web project that the Unity container uses to determine how to instantiate objects.

```csharp
C#
public static void RegisterTypes(IUnityContainer container,
  bool roleInitialization)
{
  var account = CloudConfiguration
    .GetStorageAccount("DataConnectionString");

  container.RegisterInstance(account);

  ...

  var cloudStorageAccountType =
    typeof(Microsoft.WindowsAzure.CloudStorageAccount);
  var retryPolicyFactoryProperty =
    new InjectionProperty("RetryPolicyFactory",
                          typeof(IRetryPolicyFactory));

  container
    .RegisterType<IAzureTable<SurveyRow>,
      AzureTable<SurveyRow>>(
        new InjectionConstructor(cloudStorageAccountType,
                        AzureConstants.Tables.Surveys),
      readWriteStrategyProperty,
      retryPolicyFactoryProperty)
    .RegisterType<IAzureTable<QuestionRow>,
      AzureTable<QuestionRow>>(
        new InjectionConstructor(cloudStorageAccountType,
                        AzureConstants.Tables.Questions),
      retryPolicyFactoryProperty);

  ...

  container.RegisterType<ISurveyStore, SurveyStore>
    (cacheEnabledProperty)...
}
```

When the application requires a new MVC controller instance, Unity is responsible for creating the controller. The constructor that Unity invokes to create a **SurveysController** instance takes a number of parameters including a **SurveyStore** object. The third call to the **Register-Type** method in the previous sample defines how Unity instantiates a **SurveyStore** object to pass to the **SurveysController** constructor. The first two calls to the **RegisterType** method in the previous sample define the rules that tell the Unity container how to instantiate the two **IAzureTable** instances that it must pass to the **SurveyStore** constructor shown earlier.

In the second usage scenario for the parameterized **SurveyStore** constructor, you create unit tests for the **SurveyStore** class by directly invoking the constructor and passing in mock objects created using the Moq mocking library. The following code example shows a unit test method that uses the constructor in this way.

C#

```csharp
[TestMethod]
public void GetSurveyByTenantAndSlugNameReturnsTenant
            NameFromPartitionKey()
{
  string expectedRowKey = string.Format(
    CultureInfo.InvariantCulture, "{0}_{1}", "tenant",
    "slug-name");
  var surveyRow = new SurveyRow { RowKey = expectedRowKey,
    PartitionKey = "tenant" };
  var surveyRowsForTheQuery = new[] { surveyRow };
  var mock = new Mock<IAzureTable<SurveyRow>>();
  mock.SetupGet(t => t.Query).Returns(
    surveyRowsForTheQuery.AsQueryable());
  mock.Setup(t => t.GetRetryPolicyFactoryInstance())
    .Returns(new DefaultRetryPolicyFactory());
  var store = new SurveyStore(
    mock.Object, default(IAzureTable<QuestionRow>));

  var survey = store.GetSurveyByTenantAndSlugName(
    "tenant", "slug-name", false);

  Assert.AreEqual("tenant", survey.Tenant);
}
```

The test creates a mock **IAzureTable<SurveyRow>** instance, uses it to instantiate a **SurveyStore** object, invokes the **GetSurveyByTenantAndSlugName** method, and checks the result. It performs this test without touching Windows Azure table storage.

The Surveys application uses a similar approach to enable unit testing of the other store components that use Windows Azure blob and table storage.

Testing Worker Roles

Tailspin also considered how to implement background tests in worker roles so as to minimize the effort required for unit testing. The implementation of the "plumbing" code in the worker role, and the use of Unity, makes it possible to run unit tests on the worker role components using mock objects instead of Windows Azure queues and blobs. The following code from the **BatchProcessing-QueueHandlerFixture** class shows two example unit tests.

```csharp
C#
[TestMethod]
public void ForCreatesHandlerForGivenQueue()
{
  var mockQueue = new Mock<IAzureQueue<StubMessage>>();

  var queueHandler = BatchProcessingQueueHandler
        .For(mockQueue.Object, 1);

  Assert.IsInstanceOfType(queueHandler,
        typeof(BatchMultipleQueueHandler<MessageStub>));
}

[TestMethod]
public void DoRunsGivenCommandForEachMessage()
{
  var message1 = new MessageStub();
  var message2 = new MessageStub();
  var mockQueue = new Mock<IAzureQueue<MessageStub>>();
  var queue = new Queue<IEnumerable<MessageStub>>();
  queue.Enqueue(new[] { message1, message2 });
  mockQueue.Setup(q => q.GetMessages(32))
        .Returns(() => queue.Count > 0 ?
          queue.Dequeue() : new MessageStub[] { });
  var command = new Mock<IBatchCommand<MessageStub>>();
  var queueHandler = new
      BatchProcessingQueueHandlerStub(mockQueue.Object);

  queueHandler.Do(command.Object);

  command.Verify(c => c.Run(It.IsAny<MessageStub>()),
        Times.Exactly(2));
  command.Verify(c => c.Run(message1));
  command.Verify(c => c.Run(message2));
}
```

```
public class MessageStub : AzureQueueMessage
{
}

public class CloudQueueMessageStub : CloudQueueMessage
{
  public CloudQueueMessageStub(string content)
    : base(content)
  {
    this.DequeueCount = 6;
  }
}

private class BatchProcessingQueueHandlerStub :
    BatchProcessingQueueHandler<StubMessage>
{
  public BatchProcessingQueueHandlerStub(
      IAzureQueue<StubMessage> queue) : base(queue)
  {
  }

  public override void Do(
      IBatchCommand<StubMessage> batchCommand)
  {
    this.Cycle(batchCommand);
  }
}
```

The **ForCreateHandlerForGivenQueue** unit test verifies that the static **For** method instantiates a **BatchProcessingQueueHandler** correctly by using a mock queue. The **DoRunsGivenCommand-ForEachMessage** unit test verifies that the **Do** method causes the command to be executed against every message in the queue by using mock queue and command objects.

Testing Multi-Tenant Features and Tenant Isolation

The developers at Tailspin included tests to verify that the application preserves the isolation of tenants. The following code sample shows a test in the **SurveysControllerFixture** class that verifies that the private tenant web site uses the correct tenant details when a tenant chooses to export survey data to a SQL Database instance.

```csharp
C#
[TestMethod]
public void ExportGetsTheTenantInformationAndPutsInModel()
{
  var tenant = new Tenant();

  var mockTenantStore = new Mock<ITenantStore>();
  var mockSurveyAnswerStore = new Mock<ISurveyAnswerStore>();
  mockTenantStore.Setup(
    r => r.GetTenant(It.IsAny<string>())).Returns(tenant);
  mockSurveyAnswerStore.Setup(
    r => r.GetFirstSurveyAnswerId(It.IsAny<string>(),
    It.IsAny<string>())).Returns(string.Empty);

  using (var controller = new SurveysController(
          null, mockSurveyAnswerStore.Object, null,
          mockTenantStore.Object, null))
  {
    controller.Tenant = tenant;

    var result =
      controller.ExportResponses(string.Empty) as ViewResult;

    var model = result.ViewData.Model
                as TenantPageViewData<ExportResponseModel>;

    Assert.AreSame(tenant, model.ContentModel.Tenant);
  }
}
```

Performance and Stress Testing

The test team at Tailspin conducted a set of high volume stress tests in order to determine the expected throughput with a given number of role and queue instances, and to understand how to scale the application to meet higher levels of demand. This section focuses on the specific results of stress testing the Surveys application. However, most of the factors will be relevant to the majority of Windows Azure applications.

During the stress testing exercise, the team identified a number of issues with the code that limited the scalability of the application and, as a result, the developers proposed a number of changes to overcome these limitations.

Optimistic and Pessimistic Concurrency Control

The application saves the survey summary statistics data and the list of survey responses to blob storage. A worker role collects the data, processes it, and writes it back to blob storage. When more than one worker role instance is running they could try to write to the same blob simultaneously, and so the application must use either an optimistic or a pessimistic approach to managing concurrent access issues.

As part of the stress testing, Tailspin evaluated both optimistic and pessimistic concurrency approaches when the application writes to these blobs to determine which approach enabled the highest throughput. With a heavily loaded system, and running three worker role instances, the test team saw approximately one optimistic concurrency exception per 2,000 saved survey responses. Therefore, Tailspin decided to use the optimistic concurrency approach when the application writes to these blobs.

Maintaining a List of Survey Answers

To support paging through survey answers in the order they were received by the system, and exporting to a SQL Database instance, the application maintains a list of survey responses for each survey in a blob. Chapter 3, "Choosing a Multi-Tenant Data Architecture," describes this mechanism in detail.

However, stress testing revealed that this can lead to a bottleneck in the system as the number of survey responses for a survey grows. Every time the system saves a new set of survey responses, it must read the whole list of existing responses from blob storage, append the new answers to the list, and then save the list back to blob storage.

> The results we got from our stress tests may be specific to the Surveys application, but the factors involved and our solutions are likely to be relevant to the majority of Windows Azure applications.

> Often, the only way you can make a sensible choice between optimistic and pessimistic concurrency is by testing the application to the limits, counting failures, and measuring actual performance under realistic conditions with realistic data.

The developers at Tailspin plan to address this problem by introducing a paging mechanism, so that it uses multiple blobs to store the list of survey responses for each survey. Each blob will hold a list of survey responses, but once the list reaches a certain size the application will create a new list. In this way, the size of the list that the application is currently writing to will never grow beyond a fixed size.

This will also require some changes in the logic that enables paging through survey responses in the UI and reading survey responses for export to SQL Database.

Azure Queues Throughput

According to the information in the post *"Windows Azure Storage Abstractions and their Scalability Targets"* on the Windows Azure Storage Team blog, a Windows Azure queue has a performance target of processing 500 messages per second. The Tailspin Surveys application uses two queues to deliver survey responses from the public web site to the worker role for processing (one for responses to surveys published by tenants with a standard subscription, and one for responses to surveys published by tenants with a premium subscription). It's possible, with a high volume of users responding to surveys, that the number of messages that these queues need to process could exceed 500 per second.

Tailspin plans to partition these queues, and modify the application to work with multiple instances of these queues in order to support higher rates of throughput. For example, the web role could use a round-robin approach to write messages to the multiple queue instances in turn and the worker role could use a separate thread to handle each of the queue instances. However, care is required in designing this kind of feature to ensure an appropriate number of queue instances are available when you scale the application (either manually or automatically) and the number of role instances changes.

Synchronous and Asynchronous Calls to Windows Azure Storage

The stress tests indicated that synchronously writing first to blob storage and then synchronously posting a message to a queue took up a significant portion of execution time in the web role. Typically, you can improve the throughput when you write to Windows Azure storage by using asynchronous calls to avoid blocking the application while the I/O operation completes. For example, if you need to write to storage and send a message to a queue you can initiate both operations asynchronously.

Sometimes performance bottlenecks aren't the fault of your bad code, they are limitations of services or systems you rely on. In this case you must either live with the limits, or redesign your code to find a workaround. But take care that the additional complexity you introduce does not have a greater impact on your application's performance than the limitation you originally encountered.

Just because you can do things asynchronously and concurrently doesn't always mean that you should. Some processes in an application need to be performed in a predetermined or controlled order, or must finish before the next task starts. This is particularly the case if you need to check for an error before starting the next process.

However, there are some issues that would make it difficult to convert these into asynchronous write operations in the Surveys application. For example, the web role must finish writing a survey response to blob storage before it sends a message to the queue that instructs the worker role to process it. Performing the writes to blob storage and the queue concurrently by using asynchronous code could result in errors if writing to the blob fails, or if the message arrives in the worker role before the web role finishes writing the survey response to storage.

Tailspin also considered whether it should use asynchronous calls when the application saves summary statistics and answer lists to blob storage. These write operations take place as part of the processing cycle in the worker role that consists of reading the blob data, making changes to that blob data, and then writing the data back to blob storage.

The application uses an optimistic concurrency approach that checks the data in the blob hasn't changed between the time it was read and the time that the application attempts to write it back. If the application used an asynchronous call to write the data back to blob storage, it's possible that the read operation in the next cycle will start before the previous write operation is complete—increasing the likelihood of an optimistic concurrency exception occurring.

Tailspin decided not to use asynchronous calls when the application writes summary statistics data and survey answer response lists to blob storage.

Additional Performance Tuning Options

Further performance tuning options that Tailspin will consider and test include:

- Turning off Nagling. For more information, see the post *"Nagle's Algorithm is Not Friendly towards Small Requests"* on the Windows Azure Storage Team blog.
- Setting a connection limit. For more information, see the post *"Understanding MaxServicePointIdleTime and DefaultConnectionLimit"* on the Http Client Protocol blog.
- Turning off proxy detection in the system.NET section of the web.config file when running in the cloud. See *"<proxy> Element (Network Settings)"* for details.

Managing the Surveys Application

Tailspin stores all the configuration data used to manage tenants of the Surveys application in blob storage. The private web site, defined in the Tailspin.Web project, includes a set of pages that are only available to Tailspin administrators for managing Tailspin Surveys tenants.

The sample application currently enables Tailspin administrators to add new tenants and update the details of existing tenants. It does not currently enable administrators to delete tenants.

The "Subscribers list" screen shows the Tailspin administrator a list of the current tenants in the Tailspin Surveys application. A Tailspin administrator can edit the details of existing subscribers from the subscribers list screen and add a new subscriber on the "Add a new subscriber" screen.

Tailspin plans to implement a process to enable administrators to remove a subscriber. A **Delete** hyperlink on the subscribers list screen will trigger this process, and must perform the following steps:

- Delete the tenant blob that contains the subscriber's configuration data from the **tenants** blob container.

- Delete all of the subscriber's survey questions from the **Questions** table and survey headers from the Surveys table. In the case of the **Surveys** table, each subscriber's surveys are stored on a separate partition. In the case of the **Questions** table, the partition key is a combination of the subscriber name and survey name: the delete process must find all of the partitions where the partition key starts with the subscriber's ID.

- Delete all the blob containers that contain the subscriber's survey answers (every survey has its own blob container for storing survey responses). The subscriber's ID is part of the container name.

- Delete all the blobs in the **surveyanswerssummaries** and **surveyanswerslists** blob containers that belong to the subscriber (every survey will have its own blob in each of these containers). The subscriber's ID is part of the blob names.

- Delete any data used for customizing the subscriber's surveys such as logos in the logos blob container.

- If the subscription includes a SQL Database, de-provision the database.

- Delete the subscriber's configuration data and survey definitions from the cache.

- If the subscriber uses the Tailspin identity provider, delete any accounts belonging to the subscriber from the store used by the identity provider.

Some of the actions in the previous list can be performed quickly and Tailspin plans to perform these actions synchronously when the administrator has confirmed that the subscriber must be deleted. These actions are to delete the cached data, to delete the data from the **Surveys** table, and to delete the subscriber's configuration data from the **tenants** blob container. When these items have been deleted the subscriber will not be able to access the private tenant site, and the subscriber's surveys will not be listed on the public site.

Tailspin can delete a subscriber's configuration data from blob storage quickly because the subscriber's ID is the name of the blob, it can delete the entries from the **Surveys** table quickly because all the subscriber's surveys are stored in the same partition, and it can delete the cached data quickly because the application uses a separate cache region for each tenant.

Tailspin can quickly delete some subscriber data and disable access for that subscriber. It can delete all of the remaining data later to free up storage space.

The remaining actions, which may take longer to perform, can be performed asynchronously. Deleting a subscriber's entries in the **Questions** table may take time because the entries span multiple partitions and therefore the process must scan then entire table to locate all the entries to delete. Deleting the subscriber's blobs from the **surveyanswerssummaries** and **surveyanswerslists** blob containers may take time because the process must iterate over all the blobs in the container to identify which ones belong to the subscriber.

Monitoring the Surveys Application

Tailspin uses Windows Azure diagnostics to collect information from the Surveys application at runtime. Tailspin administrators can then monitor these log files for any unexpected events or behavior. For example, the administrators can monitor the messages from the Transient Fault Handling Application Block to identify if there are any changes in Windows Azure that are affecting how the application is using Windows Azure storage or SQL Database. These types of retries will happen from time to time, which is why Tailspin uses the Transient Fault Handling Application Block. However, if the administrators see a large number of retries occurring they can take steps to investigate the status of the Windows Azure services or other dependent services.

The **AzureTable**, **AzureQueue**, and **AzureBlobContainer** classes in the application all inherit from the **AzureObjectWithRetryPolicyFactory** class that specifies the message that the application writes to the Windows Azure logs when the block detects a transient fault. The following code sample shows the **AzureObjectWithRetryPolicyFactory** class.

```C#
public abstract class AzureObjectWithRetryPolicyFactory
  : IAzureObjectWithRetryPolicyFactory
{
  public IRetryPolicyFactory RetryPolicyFactory { get; set; }

  public virtual IRetryPolicyFactory
              GetRetryPolicyFactoryInstance()
  {
    return this.RetryPolicyFactory
          ?? new DefaultRetryPolicyFactory();
  }
```

```
  protected virtual void RetryPolicyTrace(object sender,
                        RetryingEventArgs args)
  {
    var msg = string.Format(
        "Retry - Count:{0}, Delay:{1}, Exception:{2}",
        args.CurrentRetryCount,
        args.Delay,
        args.LastException);
    TraceHelper.TraceInformation(msg);
  }
}
```

ISV CONSIDERATIONS FOR MULTI-TENANT APPLICATIONS

Questions such as how to handle the onboarding process for new subscribers, how to manage per user customization, and how to implement billing are relevant to both single tenant and multi-tenant architectures. However, they require some special consideration in a multi-tenant model.

Goals and Requirements

Tailspin's goals and requirements for supporting tenants and customers that pay to use the Surveys application encompass those that are applicable to most multi-tenant applications created by ISVs.

When a new subscriber signs up for a multi-tenant application, the application must undergo configuration and other changes to enable the new account. The onboarding process must typically be automated, and it touches many components of the application. Tailspin wants to automate as much of this process as possible to simplify the onboarding process for new subscribers, and to minimize the costs associated with setting up a new subscriber.

The onboarding process touches many components in your applications.

It is common for ISVs to offer different levels of subscription, such as standard and premium subscriptions, which may vary in terms of functionality, support, and service level (for example, guaranteed availability and response times). This can make both the onboarding and the daily operation more complex to manage. Tailspin intends to offer different levels of service, and so must consider how this will affect the design of the application.

ISVs will typically want to allow tenants to customize the application, but this can add complexity to the solution and may increase security concerns if not properly controlled.

Another common feature of multi-tenant applications is enabling subscribers to customize parts of the application for their customers, such as the appearance of the UI or the availability of specific features and capabilities. The amount of customization required will vary for different scenarios and different types of application, and it is another factor that can have a large impact on the complexity of designing and managing multi-tenant applications. Tailspin intends to offer some levels of UI customization to tenants, but will limit this to simple changes such as style sheets and logos. Tailspin also wants to enable premium subscribers to add metadata, such as a product ID or an owner, to survey definitions. Premium subscribers will be able to use this contextual data as links to other data within their own systems

Finally, ISVs will need to be able to bill tenants based on their usage of the application. While Windows Azure does provide billing information for an application, calculating the costs for each tenant is less easy to achieve. Tailspin wants to be able to bill tenants at different rates based on both usage and the type of subscription that tenant has.

Overview of the Solution

This section describes the options Tailspin considered for managing individual tenants in the Surveys application, and identifies the solutions Tailspin chose.

Onboarding for Trials and New Subscribers

For Tailspin, the key issue related to onboarding is how much of the process should it automate. Building a system that handles self-service sign up is complex, but it does make it easier for potential subscribers to try out the system. The self-service onboarding process must include a number of steps, including the following:

- Validate the tenant. Tailspin must ensure that paying subscribers have a valid payment method such as a credit card.
- Create any tenant specific configuration settings. It should be possible to create (and change) tenant configuration values without restarting any part of the application. For Tailspin Surveys, tenant configuration values are stored in Windows Azure blob storage using one blob per tenant. This includes all of the information that the Tailspin federation provider needs to establish a trust relationship with the tenant's identity provider. If the tenant has chosen to use the Tailspin identity provider, the application will also need to add user accounts to the membership database. In addition, the Surveys application will use the tenant configuration data when it adds tenant identifiers to data collected at runtime by logging mechanisms, and when it performs any tenant specific backup operations.

- Provision any tenant specific resources. Tenants with premium subscriptions can choose to have their own SQL Database server to store their exported data. The SQL Database Management REST API enables you to create server instances. If you need to provision any other Windows Azure resources, such as storage accounts or cloud services, you can use the Windows Azure Service Management API.

- Notify Tailspin administrators of any additional steps that must be completed on behalf of the tenant. Tailspin does not anticipate the need for any manual steps for its administrators as part of the onboarding process.

- Notify the subscriber of any additional steps that it must take. For example, Tailspin Surveys subscribers can use a custom DNS name to access their surveys.

- Notify the subscriber of any applicable terms and conditions including the SLA for the subscription type.

> *For more information about using the Windows Azure Service Management REST APIs, see "Windows Azure Service Management REST API Reference" and "Management REST API Reference (SQL Database)."*

Configuring Subscribers

Tailspin chose to store all of the configuration data for each tenant in Windows Azure blob storage. Tailspin uses one blob per tenant and uses the JSON serializer to write the **Tenant** object to the blob. Almost all of the tenant configuration data is stored in this way, making it easy for Tailspin to manage the details of its subscribers. The only exceptions to storing tenant configuration data in blobs in the **tenants** blob container are that tenant logos are stored in the **logos** blob container, and those tenants who use the Tailspin identity provider store their users account details in the identity provider's membership database.

Supporting Per Tenant Customization

Tailspin Surveys includes three ways that subscribers can customize the application.

Each tenant can customize the UI seen by survey respondents to add tenant specific branding. Initially, each tenant will be able to upload a logo that displays on every survey page. Tailspin also plans to enable tenants to use CSS style sheets to further customize the UI. The application enables this UI customization by allowing subscribers to upload the necessary files to Windows Azure blob storage. Enabling support for custom CSS style sheets is more complex than for logos because a poorly designed style sheet could make the surveys unreadable; Tailspin plans to develop some validation and filtering functionality to minimize this risk.

We limit the types of custom CSS style selectors we accept to prevent the UI from being rendered unusable, and to protect the application from malicious attack or other unexpected side effects.

Premium tenants can add their own custom metadata to their surveys to enable linking with their own applications and services. The application uses a custom schema for each tenant to store this additional data in table storage. It also uses a custom assembly for each tenant that takes advantage of this feature, which enables the tenant to save and view this custom data in the private tenant web site. For more information about how Tailspin implemented this feature see the section "Accessing Custom Data Associated with a Survey" in Chapter 3, "Choosing a Multi-Tenant Data Architecture."

Subscribers can also customize how to authenticate with Tailspin Surveys. They can choose to use their own identity provider, Tailspin's identity provider, or a third party identity provider. This configuration data is stored in the tenant blob. For more information about how the different authentication schemes work see Chapter 6, "Securing Multi-Tenant Applications."

Financial Goals and Billing Subscribers

Tailspin developed the Surveys application as a commercial service from which it hopes to make a profit. The revenue from the application will come from tenants who sign up for one of the paid services. The costs can be broken down into the following categories:

- Tailspin incurred costs during the project to develop the Surveys application. These costs included developer salaries, software licenses, hardware, and training.
- Tailspin incurs ongoing management costs. These costs include administrator salaries, bug fixing, and developing enhancements.
- Tailspin incurs running costs. Windows Azure bills Tailspin monthly for the resources it consumes, such as web and worker role usage, data transfer, and data storage.

The costs associated with the first two categories may be difficult to identify, especially because some of the items may be associated with other projects and applications; for example, an administrator may be responsible for multiple applications. The costs in the third category are very easy for Tailspin to identify from the monthly billing statements. If the application consumes a significant quantity of Windows Azure resources, these running costs may be the most significant costs associated with the application.

The revenue that Tailspin receives from its tenants should be sufficient to generate a suitable return on investment, enabling Tailspin to recoup its initial investment costs and generate a surplus.

Tailspin evaluated two alternative pricing strategies for the Tailspin Surveys application. The first is to charge subscribers a fixed monthly amount for the package they subscribe to, the second is to charge subscribers based on their resource consumption.

Enabling tenants to extensively customize the application can add considerably to your development, test, and administration costs.

Charging subscribers a fixed monthly fee has the following advantages and disadvantages:

- Subscribers know in advance what their costs will be every month.
- Tailspin knows, based on subscriber numbers, what its income will be every month.
- There is a risk for Tailspin that, if it doesn't sign up enough subscribers, it won't cover its costs.
- For Tailspin, implementing such a billing scheme is relatively straightforward.
- It may be perceived as unfair, with some users effectively subsidizing others depending on their usage pattern.
- Tailspin must set limits that prevent subscribers from using resources excessively. With no limits in place, Tailspin may face unexpectedly large bills at the end of a month, or the performance of the application may suffer.

Charging subscribers based on their monthly resource usage has the following advantages and disadvantages:

- Tailspin can pass on its Windows Azure running costs to its tenants, plus a percentage to ensure that it always covers its monthly running costs.
- Subscribers cannot predict their monthly costs so easily.
- Subscribers may want to set a cap on their potential monthly costs, or receive notifications if they exceed a particular amount.
- Tailspin must ensure full transparency in the way that it calculates subscribers' monthly bills.
- Tailspin must add suitable monitoring to the application to accurately capture each subscriber's usage.
- This approach may be viewed as fairer because there is no cross subsidization between tenants.
- This approach is more complex to implement.

Tailspin opted for the first approach, where subscribers pay a fixed monthly fee for their subscription. Subscribers prefer this approach because their costs are predictable, and Tailspin prefers it because it can implement it relatively easily.

> *Windows Azure Marketplace can provide you with a channel for marketing your hosted service. It can also provide billing services to collect payments from subscribers. For more information, see Windows Azure Marketplace on MSDN.*

Using the Autoscaling Application Block is not just a great way to scale applications automatically—it can also be used to set upper limits on your use of cloud resources.

Tailspin will set different monthly limits for the different subscription levels. Initially, Tailspin plans to implement the following restrictions on subscribers:

- It will set different limits for premium and standard subscribers on the number of surveys they can have active at any one time. Tailspin can enforce this by checking how many surveys the subscriber currently has active whenever the subscriber tries to publish a new survey.

- It will set different limits on the duration of a survey. Tailspin can enforce this by recording, as part of the survey definition, when the subscriber published the survey. The application can check whether the maximum duration that a survey can be available for has been reached whenever it loads the list of available surveys for a subscriber.

Tailspin will also consider placing different limits on the maximum number of responses that can be collected for the different subscription levels. This will require the application to track the number of survey responses each tenant and survey receives and notify the subscriber when it is approaching the limit. The application already collects this data as part of the summary statistics it calculates.

Tailspin will monitor the application to see if any subscriber surveys result in poor performance for other users. If this occurs, it will investigate additional ways to limit the way that subscribers can consume resources.

The sample application does not currently impose any limits on the different types of subscriber.

Inside the Implementation

Now is a good time to walk through some of the code in the Tailspin Surveys application in more detail. As you go through this section, you may want to download the Visual Studio solution for the Tailspin Surveys application from *http://wag.codeplex.com/*.

Onboarding for Trials and New Subscribers

The following sections describe how Tailspin handles onboarding for new subscribers. The onboarding process collects the information described in this section and then persists it to blob storage using one blob per tenant. The web and worker roles in the Tailspin Surveys application use the tenant information in blob storage to configure the application dynamically at runtime.

Basic Subscription Information

The following table describes the basic information that every subscriber provides when they sign up for the Surveys service.

Information	Example	Notes
Subscriber Name	Adatum Ltd.	The commercial name of the subscriber. The application uses this as part of customization of the subscriber's pages on the Surveys websites. The Subscriber can also provide a corporate logo.
Subscriber Alias	adatum	A unique alias used within the application to identify the subscriber. For example, it forms part of the URL for the subscriber's web pages. The application generates a value based on the subscriber name, but it allows the subscriber to override this suggestion.
Subscription Type	Trial, Individual, Standard, Premium	The subscription type determines the feature set available to the subscriber and may affect what additional onboarding information must be collected from the subscriber.
Payment Details	Credit card details	Apart from a trial subscription, all other subscription types are paid subscriptions. The application uses a third-party solution to handle credit card payments.

Apart from credit card details, all this information is stored in Windows Azure storage; it is used throughout the onboarding process and while the subscription is active.

Authentication and Authorization Information

Chapter 6 of this guide, "Securing Multi-Tenant Applications," describes the three alternatives for managing access to the application. Each of these alternatives requires different information from the subscriber as part of the onboarding process. For example, the Standard subscription type uses a social identity provider to authenticate a user's Microsoft or Google account credentials, and the Premium subscription type can use either the subscriber's own identity provider or Tailspin's identity provider.

Provisioning a Trust Relationship with the Subscriber's Identity Provider

One of the features of the Premium subscription type is integration with the subscriber's identity provider. The onboarding process collects the information needed to configure the trust relationship between subscriber's Security Token Service (STS) and the Tailspin federation provider (FP) STS. The following table describes this information.

Information	Example	Notes
Subscriber Federation Metadata URL	https://login.adatum.net/FederationMetadata/2007-06/FederationMetadata.xml	This should be a public endpoint. An alternative is to enable the subscriber to manually upload this data.
Administrator identifier (email or Security Account Manager Account Name)	john@adatum.com	The Surveys application creates a rule in its FP to map this identifier to the administrator role in the Surveys application.
User identifier claim type	http://schemas.xmlsoap.org/ws/2005/05/identity/claims/name	This is the claim type that the subscriber's STS will issue to identify a user.
Thumbprint of subscriber's token signing key	d2316c731b39683b743109278c81e2684523d17e	The federation provider STS compares this to the thumbprint of the certificate included in the security token sent by the subscriber's STS. If they match, the Tailspin federation provider can trust the security token.
Claims transformation rules	Group:Domain Users => Role:Survey Creator	These rules map a subscriber's claim types to claim types understood by the Surveys application.

The sample code includes the Tailspin.SimulatedIssuer project, which includes a simple federation provider that manages the federation with Tailspin's subscribers. This federation provider reads the information it needs from the tenant's configuration data in blob storage. The following code sample from the **FederationSecurityTokenService** class in the Tailspin.SimulatedIssuer project shows how this simple federation provider uses the tenant information to perform the claims transformation from the tenant's claim into a claim that the Tailspin Surveys application recognizes.

```C#
protected override IClaimsIdentity GetOutputClaimsIdentity(
        IClaimsPrincipal principal,
        RequestSecurityToken request,
        Scope scope)
{
  if (principal == null)
  {
    throw new InvalidRequestException(
      "The caller's principal is null.");
  }

  var input = principal.Identity as ClaimsIdentity;

  var tenant = this.tenantStore.GetTenant(
                        input.Claims.First().Issuer);
  if (tenant == null)
  {
    throw new InvalidOperationException(
      "Issuer not trusted.");
  }

  var output = new ClaimsIdentity();

  CopyClaims(input,
            new[] { WSIdentityConstants.ClaimTypes.Name },
            output);
  TransformClaims(input, tenant.ClaimType,
                tenant.ClaimValue, ClaimTypes.Role,
                Tailspin.Roles.SurveyAdministrator,
                output);
  output.Claims.Add(new Claim(Tailspin.ClaimTypes.Tenant,
                            tenant.Name));

  return output;
}
```

The following code sample from the **TenantStoreBasedIssuerNameRegistry** in the Tailspin.Simulated-Issuer project shows how the Tailspin federation provider verifies that a security token is from a trusted source. It compares the subscriber's thumbprint stored in the tenant configuration data with the thumbprint of the signing certificate in the security token received from the tenant's STS.

```csharp
C#
public override string GetIssuerName(
                        SecurityToken securityToken)
{
  if (securityToken is X509SecurityToken)
  {
    string thumbprint = (securityToken as X509SecurityToken)
                        .Certificate.Thumbprint;
    foreach (
      var tenantName in this.tenantStore.GetTenantNames())
    {
      var tenant = this.tenantStore.GetTenant(tenantName);
      if (tenant.IssuerThumbPrint.Equals(thumbprint,
        System.StringComparison.InvariantCultureIgnoreCase))
      {
        return tenant.Name;
      }
    }
    return null;
  }
  else
  {
    throw new InvalidSecurityTokenException(
              "Empty or wrong securityToken argument");
  }
}
```

In the future Tailspin could decide to use ADFS, Windows Azure Access Control, or a different custom STS as its federation provider STS. As part of the onboarding process, the Surveys application would have to programmatically create the trust relationship between the Tailspin federation provider STS and the subscriber's identity provider, and programmatically add any claims transformation rules to the Tailspin federation provider.

> *For more information about using claims and trust relationships see the section "Setup and Physical Deployment" in Chapter 5, "Federated Identity with Windows Azure Access Control Service," of "A Guide to Claims-Based Identity and Access Control."*

Provisioning Authentication and Authorization for Basic Subscribers

Subscribers to the Standard subscription type cannot integrate the Surveys application with their own STS. Instead, they define their own users in the Surveys application. During the onboarding process they provide details for the administrator account that will have full access to everything in their account, including billing information. They can later define additional users as members of the Survey Creator role, who can only create surveys and analyze the results.

Provisioning Authentication and Authorization for Individual Subscribers

Individual subscribers use a third-party social identity such as a Microsoft account, Open ID credentials, or Google ID credentials to authenticate with the Surveys application. During the onboarding process they must provide details of the identity they will use. This identity has administrator rights for the account and is the only identity that can be used to access the account.

Geo-location Information

You could automatically suggest a location based on the user's IP address by using a service such as the IPInfoDB IP Location XML API.

During the onboarding process, the subscriber selects the geographic location where the Surveys application will host its account. The list of available locations is a subset, chosen by Tailspin, of the locations where there are currently Windows Azure data centers. This geographic location identifies the location of the Subscriber website instance that the subscriber will use, and where the application stores data associated with the account. It is also the default location for hosting the subscriber's surveys, although the subscriber can opt to host individual surveys in alternate geographical locations. For more information about how Tailspin plans to implement this behavior, see Chapter 5, "Maximizing Availability, Scalability, and Elasticity." Currently, the sample application allows a subscriber to select a hosting location, saves this in the tenant configuration, but does not use it.

Database Information

During the sign-up process, a subscriber can also opt to provision a Windows Azure SQL Database instance to store and analyze its survey data. The application creates this database on a SQL Database server in the same geographical location as the subscriber's account. The application uses the subscriber alias to generate the database name and the database user name. The application also generates a random password. The application saves the database connection string in Windows Azure storage, together with the other subscriber account data.

At the time of writing, there is a soft limit of 150 databases per SQL Database server. Tailspin could monitor manually how many databases are created on each SQL Database server, and then add new server instances as required. Alternatively, Tailspin could automate this process using the SQL Database Management REST API. For more information, see *"Operations on Windows Azure SQL Database Servers."*

> *The Windows Azure SQL Database instance is owned and paid for by Tailspin. Tailspin charges subscribers for this service. For more information about how the Surveys application uses Windows Azure SQL Database see the section "Implementing the Data Export" in Chapter 3, "Choosing a Multi-Tenant Data Architecture," of this guide.*

Customizing the Surveys Application for Each Subscriber

A common feature of multi-tenant applications is enabling subscribers to customize features of the application for their subscribers, such as the appearance of the application and the availability of selected UI features and functionality.

How Tailspin Allows Subscribers to Customize the User Interface

The current version of the Surveys application enables subscribers to customize the appearance of their pages by using a custom logo image. Subscribers can upload an image to their account, and the Surveys application saves the image as part of the subscriber's account data in blob storage. The application can then display the image on pages in the public and private web sites.

The current solution allows a subscriber to upload a single image to a public blob container named **logos**. As part of the upload process, the application adds the URL for the logo image to the tenant's blob data stored in the blob container named **tenants**. The following code sample from the **Tenant-Store** class shows how the application saves the subscriber's logo image to blob store and then updates the tenant's configuration data with the URL of the image:

```C#
public void UploadLogo(string tenant, byte[] logo)
{
  this.logosBlobContainer.Save(tenant, logo);

  var tenantToUpdate =
        this.tenantBlobContainer.Get(tenant);
  tenantToUpdate.Logo =
        this.logosBlobContainer.GetUri(tenant).ToString();

  this.SaveTenant(tenantToUpdate);
}
```

Tailspin plans to extend the customization options available to subscribers in future versions of the application. These planned extensions, which are not included in the sample, will enable subscribers to customize the appearance of their survey pages to follow corporate branding by using cascading style sheets (CSS) technology.

Tailspin is concerned about the security implications of allowing subscribers to upload custom .css files, and plans to limit the CSS features that the site will support. To do this, Tailspin plans to provide a UI where subscribers can provide custom definitions for a predefined list of CSS selectors that are applied to the HTML elements used to display the survey page and its questions. The Surveys application will store these custom CSS selector definitions as part of each tenant's configuration data, enabling each subscriber to customize its surveys using its own style. The following code sample shows a selection of the CSS selectors that the application currently uses and that could, potentially, be overridden using this approach.

```CSS
#surveyTitle
{
    ...
}
#surveyTitle h1
{
    ...
}

#surveyForm
{
    ...
}

#surveyForm ol
{
    ...
}

#surveyForm ol li
{
    ...
}

#surveyForm .option input[type="radio"]
{
    ...
}

.stars span span
{
    ...
}

.stars span.rating-over
{
    ...
}

.stars span.rating
{
    ...
}
```

The Surveys application will construct a custom style sheet dynamically at runtime using the custom definitions saved by the subscriber, and link to it in the HTML pages. The following code sample shows how the Survey Display page in the public site might apply the custom CSS selectors defined by the Adatum subscriber.

```HTML
<head>
    <meta http-equiv="Content-Type" content="text/html;
        charset=utf-8" />
    <meta http-equiv="X-UA-Compatible" content="IE=8" />
    <title>Tailspin - Survey #1</title>
    <link href="/Content/styles/baseStyle.css"
        rel="stylesheet" type="text/css" media="screen" />
    <link href="/Utility/DynamicStyle.aspx?TenantID=adatum"
        rel="stylesheet" type="text/css" media="screen" />

</head>
```

The page imports the custom styles generated by the DynamicStyle. aspx page after the default styles so that any customizations defined by the subscriber override the base styles.

Tailspin will implement a scanning mechanism to verify that the CSS customizations provided by the tenants do not include any of the CSS features that the Surveys site does not support, or that could compromise the application's security.

Cascading style sheets behaviors are one feature that the Surveys site will not support.

Billing Subscribers in the Surveys Application

Tailspin plans to bill each subscriber a fixed monthly fee to use the Surveys application. Subscribers will be able to subscribe to one of several packages, such as those outlined in the following table.

Subscription type	User accounts	Maximum survey duration	Maximum active surveys
Trial	A single user account linked to a social identity provider, such as Windows Live or OpenID.	5 days	1
Basic	A single user account linked to a social identity provider, such as Windows Live or OpenID.	14 days	1
Standard	Up to five user accounts provided by the Surveys application.	28 days	10
Premium	Unlimited user accounts linked from the subscriber's own identity provider.	56 days	20

The advantage of this approach is simplicity for both Tailspin and the subscribers, because the monthly charge is fixed for each subscriber. Tailspin must undertake some market research to estimate the number of monthly subscribers at each level so that it can set appropriate charges for each subscription level.

In the future Tailspin wants to be able to offer extensions to the basic subscription types. For example, Tailspin wants to enable subscribers to extend the duration of a survey beyond the current maximum, or increase the number of active surveys beyond the current maximum. To do this, Tailspin will need to be able to capture usage metrics from the application to help it calculate any additional charges incurred by a subscriber.

Tailspin must have good estimates of expected usage to be able to estimate costs, revenue, and profit.

At the time of writing, the best approach to capturing usage metrics is via logging. Several log files are useful. You can use the Internet Information Services (IIS) logs to determine which tenant generated the web role traffic. Your application can write custom messages to the WADLogsTable in response to events such as a survey being completed. The sys.bandwidth_usage view in the master database of each Windows Azure SQL Database server shows bandwidth consumption by database.

More Information

All links in this book are accessible from the book's online bibliography available at: *http://msdn.microsoft.com/library/jj871057.aspx*.

For more information about ALM and Windows Azure, see the articles listed at *"Testing, Managing, Monitoring and Optimizing Windows Azure Applications"* on MSDN.

For information about creating custom performance counters, see *"Real World: Creating Custom Performance Counters for Windows Azure Applications with PowerShell."*

For a discussion of how to ensure business continuity with Windows Azure applications, see *"Business Continuity for Windows Azure."*

For information about the SQL Database Import/Export Service, see *"How to: Import and Export a Database (Windows Azure SQL Database)."*

For a useful collection of links and resources related to testing Windows Azure applications, see *"Testing Applications in Windows Azure."*

For more information about monitoring your Windows Azure application, including how to use Microsoft System Center Operations Manager, see *"Troubleshooting in Windows Azure."*

For information about the differences between the local compute and storage emulators and the Windows Azure services, see *Differences Between the Storage Emulator and Windows Azure Storage Services* and *Differences Between the Compute Emulator and Windows Azure.*

Glossary

affinity group. A named grouping that is in a single data center. It can include all the components associated with an application, such as storage, Windows Azure SQL Database instances, and roles.

ASP.NET MVC. A framework for developing web applications. It is based on the Model-View-Controller architectural design pattern.

autoscaling. Automatically scaling an application based on a schedule or on metrics collected from the environment.

claim. A statement about a subject; for example, a name, identity, key, group, permission, or capability made by one subject about itself or another subject. Claims are given one or more values and then packaged in security tokens that are distributed by the issuer.

cloud. A set of interconnected servers located in one or more data centers.

cloud service. Windows Azure environment where you host your application's web and worker roles. Formally referred to as a hosted service.

code near. When an application and its associated database(s) are both in the cloud.

code far. When an application is on-premises and its associated database(s) are in the cloud.

compute emulator. The Windows Azure compute emulator enables you to run, test, debug, and fine-tune your application before you deploy it as a hosted service to Windows Azure. See also: storage emulator.

Content Delivery Network (CDN). A system composed of multiple servers that contain copies of data. These servers are located in different geographical areas so that users can access the copy that is closest to them.

continuation token. A technique, supported by Windows Azure table storage, which enables a client to page through records. In response to a query, a server returns a page of records and a continuation token. If the client submits the continuation token back to the server, the server delivers the next page of records.

elasticity. A property of a system that describes its ability to scale in and out dynamically.

Enterprise Library. A collection of reusable software components (application blocks) designed to assist software developers with common enterprise development cross-cutting concerns (such as logging, validation, data access, exception handling, and many others).

entity group transaction (EGT). A transaction with ACID properties across multiple entities stored in the same Windows Azure table partition.

federation. In Windows Azure SQL Database, a federation is a way to scale out horizontally by using additional servers. Also known as sharding.

federation provider. A special case of a Security Token Service (STS) that typically trusts a third-party identity provider. The federation provider may transform the claims in the token from the third-party identity provider into a format acceptable to your application.

horizontal scalability. The ability to add more servers that are copies of existing servers.

hosted service. Spaces where applications are deployed.

idempotent operation. An operation that can be performed multiple times without changing the result. An example is setting a variable.

identity provider. Typically, a separate system that is responsible for determining the identity of a user. An application trusts an identity provider to perform this task. The identity provider passes information about the user in the form of a token. An identity provider is a special case of a Security Token Service (STS).

Infrastructure as a Service (IaaS). A collection of infrastructure services such as storage, computing resources, and network that you can rent from an external partner.

lease. An exclusive write lock on a blob that lasts until the lease expires.

mock. A mock object is used in a test to simulate a real object. They are useful when it is impractical to use the real object in the test.

optimistic concurrency. A concurrency control method that assumes that multiple changes to data can complete without affecting each other; therefore, there is no need to lock the data resources. Optimistic concurrency assumes that concurrency violations occur infrequently and simply disallows any updates or deletions that cause a concurrency violation.

Platform as a Service (Paas). A collection of platform services that you can rent from an external partner that enable you to deploy and run your application without the need to manage any infrastructure.

poison message. A message that contains malformed data that causes the queue processor to throw an exception. The result is that the message isn't processed, stays in the queue, and the next attempt to process it once again fails.

Representational State Transfer (REST). An architectural style for retrieving information from websites. A resource is the source of specific information. Each resource is identified by a global identifier, such as a Uniform Resource Identifier (URI) in HTTP. The representation is the actual document that conveys the information.

role. A web or worker role to deploy to Windows Azure.

role instance. A running instance of a web or worker role in Windows Azure.

secure sockets layer (SSL). A cryptographic protocol that uses public key cryptography to secure communication over the internet, for example using the HTTPS protocol.

Security Token Service (STS). A service that issues claims in the form of tokens. An application may be configured to trust the tokens issued by a specific STS.

service configuration file. Sets values for the service that can be configured while the hosted service is running. The values you can specify in the service configuration file include the number of instances that you want to deploy for each role, the values for the configuration parameters that you established in the service definition file, and the thumbprints for any SSL certificates associated with the service.

service definition file. Defines the roles that comprise a service, optional local storage resources, configuration settings, and certificates for SSL endpoints.

service level agreement (SLA). The formal definition of the level of service that a service provider undertakes to deliver to the customer. For example, specifying the number of hours that a service will be available for every month.

service package. Packages the role binaries and service definition file for publication to the Windows Azure Cloud Services.

sharding. See federation.

shared access signatures (SAS). A special URL that can be used to gain temporary access to data in Windows Azure table, queue, blob, and blob storage. By generating a SAS URL and giving it to a client, you can grant the client temporary and limited access to data.

snapshot. A read-only copy of a blob.

Storage Emulator. The Windows Azure storage emulator provides local instances of the blob, queue, and table services that are available in Windows Azure. If you are building an application that uses storage services, you can test locally by using the storage emulator.

throttling. The behavior of a Windows Azure service when it restricts the throughput from one client application in order to ensure that other client applications can continue to use the service.

transient faults. Error conditions that can occur in a distributed environment and that often disappear when you retry the operation. These are often caused by transient problems with the network.

vertical scalability. The ability to increase a computer's resources, such as memory or CPUs.

Web role. An interactive application that runs in the Windows Azure environment. A web role can be implemented with any technology that works with Internet Information Services (IIS) 7. See Windows Azure Cloud Services.

Windows Azure. Microsoft's platform for cloud-based computing. It is provided as a service over the Internet using either the PaaS or IaaS approaches. It includes a computing environment, the ability to run virtual machines, Windows Azure storage, and management services.

Windows Azure Cloud Services. Web and worker roles in the Windows Azure environment that enable you to adopt the PaaS approach.

Windows Azure Management Portal. A web-based administrative console for creating and managing your Windows Azure hosted services, including Cloud Services, SQL Database, storage, Virtual Machines, Virtual Networks, and Web Sites.

Windows Azure SQL Database. A relational database management system (RDBMS) in the cloud. Windows Azure SQL Database is independent of the storage that is a part of Windows Azure. It is based on SQL Server and can store structured, semi-structured, and unstructured data.

Windows Azure storage. Consists of blobs, tables, drives, and queues. It is accessible with HTTP/HTTPS requests. It is distinct from Windows Azure SQL Database.

Windows Azure Traffic Manager. A Windows Azure service that enables you to control how Windows Azure routes traffic to your cloud services.

Windows Azure Virtual Machine. Virtual machines in the Windows Azure environment that enable you to adopt the IaaS approach.

Windows Azure Virtual Network. Windows Azure service that enables you to create secure site-to-site connectivity, as well as protected private virtual networks in the cloud.

Windows Azure Web Sites. A Windows Azure service that enables you to quickly and easily deploy websites that use client and server side scripting and databases to the cloud.

Worker role. Performs batch processes and background tasks in the Windows Azure environment. Worker roles can make outbound calls and open endpoints for incoming calls. Worker roles typically use queues to communicate with Web roles. See Windows Azure Cloud Services.

Index

A

acknowledgments, xviii-xxi

applications

    architecture, 14

    authentication and authorization, 19

    code bases, 20

    costs management, 26

    CQRS pattern, 20

    customizing, 22-23

    engineering costs, 26

    financial considerations, 24-26

    fixed monthly fee plans, 25-26

    geo-location, 19

    legal and regulatory environment, 19

    life cycle management, 20-22

    monitoring, 21

    multiple multi-tenant instances, 17

    pay-per-use plans, 25

    provider's perspective, 10-11

    resource limitations and throttling, 18

    scalability, 15-18

    SLAs, 19

    stability, 14-15

    tenant's perspective, 9-10

    third-party components, 21

    trials and new subscribers, 22

    updates, 21

    URL schemes, 23-24

    vs. single-tenant model, 11-12

    Windows Azure, 9-27

architecture, 29-69

    applications, 14

    comparing paging solutions, 52-53

    custom data associated with a survey, 56-62

    custom field writing to the Surveys table, 57-61

    custom fields from the Surveys table, 61-62

    data architectures, 32-42

    data architectures scalability, 38-42

    data export implementing, 64-66

    Display method, 67

    exporting survey data to SQL Database, 43-44

    extensibility, 36-38, 43

    goals and requirements, 42-44

    Html.DisplayFor element, 68

    Html.EditorFor element, 67

    inside the implementation, 55-68

    new custom fields, 48

    ordered list of Survey responses, 62

    paging implementation, 62-64

    paging through survey results, 43

    paging with blob storage, 53

    paging with table storage, 52-53

    partitioning to isolate tenant data, 32-35

    questions display, 66-67

Questions table, 47

saving user-defined fields in a new survey, 58

scalability, 43

shared access signatures, 35

solution overview, 44-54

SQL Database design, 53-54

storage accounts, 44

storage availability, 31-32

store classes, 55-56

summary statistics displaying, 68

Survey answer storing, 50-51

Survey answer summaries, 51-52

survey definition storing, 45-48

SurveyAnswer object, 50

SurveyAnswersSummaryStore class, 55

SurveyAnswerStore class, 55

Surveys data model, 44-52

Surveys table, 46

Surveys table structure in Windows Azure SQL Database, 54

SurveysController class, 63-64, 67

SurveySqlStore class, 55, 65

SurveyStore class, 55

SurveyTransferMessage class, 64-65

SurveyTransferStore class, 55

tenant's custom fields, 56-57

tenant's data isolation, 42

tenant's data storing, 49

TenantStore class, 56

Windows Azure blob storage, 30

Windows Azure data storage, 29-32

Windows Azure SQL Database, 30-31

Windows Azure table storage, 29-30

audience, xiii

authentication and authorization, 19

See also security

automation, 11

availability, 10

availability, scalability, and elasticity, 113-156

access control for the blob containers, 141

Autoscaling Application Block, 147

availability in multi-tenant applications, 113-114

background task types, 121

blob vs. table storage, 136

caching, 115

caching policy, 143

CDN configuring and storing the content, 141-142

configuring URLs to access the content, 142

Content Delivery Network (CDN), 116, 140-141

delayed write pattern, 128-129

elasticity, 115, 126

execution model, 120-121

geo-location, 125-126

goals and requirements, 123-126

impact on other parts of the system, 135-136

inside the implementation, 147-156

large messages, 130-131

MapReduce algorithm, 123

minimizing storage transactions, 135

multiple worker role instances, 122

option comparison, 132-134

pessimistic and optimistic concurrency handling, 154-155

saving response data, 123-124

saving the response data asynchronously, 148-150

scalability, 126

scalability in multi-tenant applications, 114-116

Shared Access Signatures (SAS), 116

solution overview, 127-147

SQL Database Federation, 115

summary statistics, 124-125

summary statistics calculation, 150-154

summary statistics options, 137-138

summary statistics scalability, 139

survey response saving options, 127-136

synchronizing survey statistics, 145-146

Tailspin surveys in multiple locations, 144-145

triggers background tasks, 119-120

UI responsiveness when saving survey responses, 134

Windows Azure applications with worker roles, 117-123

Windows Azure Caching, 139

worker role scenarios, 118-119

worker role tasks, 131-132

writing directly to storage, 127-128

B

Bharath *See* cloud specialist role (Bharath)

billing, 10

C

cloud specialist role (Bharath), xvi

code bases, 20

costs, 10

  management, 26

CQRS pattern, 20

custom data, 56-62

custom fields

  new, 48

  to the Surveys table, 57-61

  from the Surveys table, 61-62

customizability, 10

D

data architecture, 32-42

  scalability, 38-42

data export implementing, 64-66

Display method, 67

E

elasticity *See* availability, scalability, and elasticity

engineering costs, 26

extensibility, 36-38, 43

F

financial considerations, 24-26

fixed monthly fee plans, 25-26

foreword, xi

G

geo-location, 19

glossary, 215-218

goals and requirements, 42-44

guide structure, xiv-xv

H

Html.DisplayFor element, 68

Html.EditorFor element, 67

I

isolation, 9

IT professional role (Poe), xvii

J

Jana *See* software architect role (Jana)

L

legal and regulatory environment, 19

life cycle management, 20-22

M

maintainability, 11

managing and monitoring, 177-213

  additional performance tuning options, 196

  ALM considerations, 177-199

  application management strategies, 182-185

  application monitoring strategies, 185-186

  authentication and authorization

basic subscribers, 208

individual subscribers, 208

information, 205

Azure queues throughput, 195

AzureObjectWithRetryPolicyFactory
class, 198-199

AzureTable class, 186-188

backup and restore for data, 184-185

basic subscription information, 204-205

BatchProcessingQueueHandlerFixture
class, 191-192

ContainerBootstrapper class, 189

CSS style sheets, 201

database information, 208-209

deployment and update strategies, 182

FederationSecurityTokenService class, 206

financial goals and billing subscribers, 202-203

geo-location information, 208

goals and requirements, 177-179

IAzureTable interface, 186-188

inside the implementation, 186-199, 204-212

ISV considerations, 199-212

multi-tenant features and tenant isolation, 193

onboarding for trials and new subscribers, 204

optimistic and pessimistic concurrency control, 194

per tenant customization, 201-202

performance and stress testing, 194-196

performance tuning, 181

reliability and availability, 183

solution overview, 200-204

stress testing, 181

subscriber billing, 212

subscriber configuring, 201

survey answers list, 194-195

Surveys application

customizing, 209-211

managing, 197-198

monitoring, 198-199

SurveysControllerFixture class, 193

SurveyStore class, 188, 190

synchronous and asynchronous calls to Windows
Azure storage, 195-196

TenantStoreBasedIssuerNameRegistry class, 207

testing strategies, 179-181

trust relationship with the subscriber's identity
provider, 205-207

UI customizing, 209-211

unit testing, 186-191

unit testing support, 180-181

worker roles testing, 191-192

Markus *See* senior software developer role (Markus)

monitoring, 11

applications, 21

more information, xvi

multi-tenant applications *See* applications

multi-tenant applications partitioning *See* partitioning

multi-tenant applications securing *See* security

multi-tenant architecture *See* architecture

multiple multi-tenant instances, 17

multiple service levels, 11

N
new custom fields, 48

O
ordered list of Survey responses, 62

P
paging

with blob storage, 53

implementation, 62-64

solutions, 52-53

with table storage, 52-53

through survey results, 43

partitioning, 71-111

AppRoutes class, 99

BatchMultipleQueueHandler class, 92-97

cached tenant data, 89-90

caches, 80-81

caching frequently used data, 108-111

cost, 88

DNS names, certificates, and SSL in the Surveys application, 85-87

goals and requirements, 81-83

http://tailspin.cloudapp.net, 87

https://tailspin.cloudapp.net, 86

identifying the tenant in a web role, 74-78

identifying the tenant in a worker role, 77-78

inside the implementation, 90-111

isolate tenant data, 32-35

isolation, 81

key plumbing types, 92

multi-instance, multi-tenant mode, 71

multi-instance, single-tenant model, 71

MVC routing tables, 97-100

performance, 89

premium subscriptions, 82

prioritizing work in a worker role, 90-97

queues, 78-80

queues and worker roles partitioning, 84

RegisterArea method, 100

robustness, 89

scalability, 81-82, 89

security, 89

ServiceDefinition.csdef file, 101

session management, 102-108

session state, 87-88

session state provider in the TailSpin.Web application, 107-108

simplicity, 88

single instance, multi-tenant model, 71

solution overview, 84-90

SSL, 76

survey designing, 83

Surveys application accessing, 82

surveys in different regions, 87

tenant isolation in web roles, 84-85

TenantCacheHelper class, 108-111

user experience, 89

web or worker role partitioning, 73-78

web roles in Tailspin surveys, 100-102

Windows Azure application partitioning, 71-81

Windows Azure Caching cache configuring, 106-107

pay-per-use plans, 25

Poe *See* IT professional role (Poe)

preface, xiii-xvii

profitability, 10

provisioning, 11

Q

questions display, 66-67

Questions table, 47

R

regulatory compliance, 10

relevance, xiv

requirements, xv-xvi

resources, xvi

limitations and throttling, 18

roles *See* cloud specialist role (Bharath); IT professional role (Poe); senior software developer role (Markus); software architect role (Jana); who's who

S

scalability, 10

See also availability, scalability, and elasticity

applications, 15-18

architecture, 43

security, 157-175

authentication, 157, 163

authorization, 158, 163

goals and requirements, 163

identity federation for tenants, 168

identity mechanism for small organizations, 165-166

privacy, 163

sensitive data, 158-159

session token protecting in Windows Azure, 174-175

session tokens encrypting in a Windows Azure application, 169

Shared Access Signatures (SAS), 161-163

social identity providers, 166-167

splitting sensitive data across multiple subscriptions, 160-161

subscriber's own identity mechanism, 164-165

Surveys application scenarios, 164-168

Windows Azure Access Control, 167

Windows Azure Active Directory, 167

Windows Identity Foundation (WIF), 170-174

security *See* multi-tenant applications securing

senior software developer role (Markus), xvii

shared access signatures, 35

single-tenant model, 11-12

SLAs, 19

software architect role (Jana), xvii

SQL Database design, 53-54

SQL Database Federation, 115

stability, 14-15

storage

accounts, 44

availability, 31-32

store classes, 55-56

structure, xiv-xv

summary statistics displaying, 68

survey data export to SQL Database, 43-44

survey definition storing, 45-48

SurveyAnswer object, 50

SurveyAnswersSummaryStore class, 55

SurveyAnswerStore class, 55

Surveys application, 5

answer storing, 50-51

answer summaries, 51-52

described, 2-3

Surveys data model, 44-52

Surveys table, 46

Surveys table structure in Windows Azure SQL Database, 54

SurveysController class, 63-64, 67

SurveySqlStore class, 55, 65

SurveyStore class, 55

SurveyTransferMessage class, 64-65

SurveyTransferStore class, 55

system requirements, xv-xvi

T

Tailspin scenario, 1-7

topic areas, 6

target audience, xiii

tenants

custom fields, 56-57

data isolation, 42

data storing, 49

perspective, 9-10

TenantStore class, 56

terminology, 215-218

third-party components, 21

trials and new subscribers, 22

U

updates, 21

URL schemes, 23-24

user-defined fields in a new survey, 58

W

who's who, xvi-xvii

Windows Azure

applications, 9-27

blob storage, 30

data storage, 29-32

Windows Azure SQL Database, 30-31